STRUCTURAL ENGINEERING ART AND APPROXIMATION

Dedicated to all those engineers, architects, designers, contractors and clients who have helped me to design aesthetically pleasing and practical structures.

www.struartapp.com

First Edition 2014
Second Edition 2015
Third Edition 2016
Fourth Edition 2017
Fifth Edition 2019

Print Edition ISBN 978-1-917336-86-4

Original page Layout Design by Amanda Hancock
Cover photograph by David Laslett of Manchestercreativemedia
Cover Design David Colon of fiverr

Acknowledgements

This book has taken more than three years to put together, having gone through numerous alterations. Over this time, much has been learnt about book editing, producing diagrams from CAD, which are readable on tablets, and just how much that the writer, despite thirty years of experience, can still make schoolboy errors with simple calculations and the use of consistent units.

Firstly, I must thank Amanda Hancock and Henry Wong for their expertise in graphic design, without whom I would not have learned so much about laying out a book. They have guided me through the process and have assisted with laying out pages, presenting images, and avoiding 'orphans,' commas, and hyphenations. They have taken me through legibility, consistency, and clarity, exactly what the book is about.

David Laslett, of Manchestercreativemedia, has taken professional photographs for the book cover. He is currently designing the website to accompany this book. A pleasure to work with and a font of knowledge in photography and web design.

Dr David Brohn has offered useful insight and comments into the book along with encouragement which is greatly appreciated. His work in teaching engineers qualitative analysis has been well-received in the profession. He is very keen on the message in this book being widely spread throughout the profession.

David Yeomans gave me publishing advice and moral support with my endeavours as an experienced writer of engineering textbooks, all without prejudice.

Dr Aleksander Pavic of Exeter University, a world authority on the dynamics of floors, very kindly offered to review **chapter 2.12** on dynamics. I hasten to add that my initial attempts contained many misunderstandings of the principles of this subject. I may also have presented, in his view, a somewhat crude and outdated introduction to the subject. I have gone ahead, nevertheless, and would not hold him in any way responsible for anything that I have presented in this book.

My friend, Florian Foerster, reviewed a previous version of this book. He did not hold back in his comments, which were appreciated. It is much more consistent as a result of this.

My former employer, Arup, has been very helpful. Bruce Danzinger in the California office provided an image for the Peter Rice Seville Arches. Thomas Li from the London Office was kind enough to comment on my efforts with fabric roofs.

Edwin Tusubira, a graduate engineer at hma for a time, was kind enough to offer his comments from a graduate point of view. I value his honesty and enthusiasm for the project.

My heartfelt thanks go to my father, John Morrison, who proofread some of the chapters for me at an early stage. His comments were, in particular, most gratefully received. He has never ceased to amaze me just how much he knows about diverse subjects and engineering and historic architecture.

My wife, Bernadette, has put up with this project for a good three years now. I have bored her to tears with the ins and outs of a project that may never go any further than this proof. It is on those one loves most that one places the most trust that they will put up with various hair-brained projects.

Finally, none of this knowledge would have been gained without my engineering mentors. I have been privileged to have worked with and for in my formative years. They constitute some of the best professionals in the art of structural engineering: Steven Morley (formerly at Modus), Chris Wise (Expedition), Ian Donaldson (late, formerly Modus), John Melling (Arup Associates), Rob Wallis (Arup), Andy McNulty (Arup), Tony Stevens (Ove Arup), Charles King (BAe/SCI). There are so many more, some of whom will have completely forgotten that I exist.

Last, but by no means least, to Mark Edwards for checking the chapter about foundation design, with whom I have had many discussions about the need for approximate methods.

Never forgotten will be that short period when I joined Arup Associates and enjoyed, all of a sudden, a breath of fresh air: architects and engineers enjoying working in a room together as true collaborators, with no particular demarcation of roles or hierarchy, in short to design good buildings. I thank them for having enticed me back into the profession after a three-year stint as a stress engineer in the aerospace industry.

Structural Engineering Art and Approximation

Contents

Part 1 Introduction

Part 2 A Visual Approach to Design with Examples

Part 3 A Review of Failures and Practice

Prologue

There are textbooks that cover structural analysis, others that tackle design, and some that are comprehensive references for areas of study such as stress analysis. There are few guides that are written from the viewpoint of practising engineers or which are there to train engineers in the art of design; which take structural problems of everyday engineering; to question, for instance: 'what is useful?', 'why do engineers adopt this approach?' ,'when is this appropriate?' or 'where is this used?'. There are few textbooks that minimise complicated mathematics and design standards, offering an introduction to structural design, which does not intimidate.

This book is intended for undergraduates in their second year onwards through to newly graduated engineers and architects. It gathers together conceptual design methods used by practising engineers for those who seek a clear guide to generic structural problems. In a busy design office, many books and papers are not readily accessible in the sense that puzzling over equations and opaque logic offers little but further obscurity. Materials are reviewed and their appropriateness for particular uses, at the same time, avoiding reference to codes in a manner that aims to bring clarity to the vexed art of structural design.

From the simplest structures, calculation methods are slowly introduced. A qualitative approach with historical references moves into quantitative methods in the later chapters. Units are introduced gradually. The intention is to invite the reader to appreciate that structural design is an art in which the tools of the trade are analytical techniques. It is intended to remove the fear of maths and methodology; to appreciate that even the most breath-taking structures may be conceived with simple principles.

By using simple techniques at the concept stage, a design may be developed with confidence that the characteristics of a particular structure are understood rather than surmised.

In the first part of the new millennium, a dazzling array of materials is available: glulam timber, carbon fibre, aluminium, glass structures, kevlar, nylon, glass fibre, and others. Methods of analysis are undertaken, which would be barely possible without the computer, for example, finite element analysis, time history dynamic analysis, and non- linear plastic analysis of plates.

Design starts at the bottom of a pyramid of choices: selection of materials, method of analysis, structural form, and appropriateness. This all takes place before any detailed design commences. At the top of the pyramid, the final design entails a process of eliminating inappropriate solutions. A contingent part of this procedure will involve sketches and approximate calculations.

Every undergraduate now has their own lap top or personal computer with which they can analyse very complicated structures, increasingly using hand held tablets and even mobile phones with apps. But regardless of technology, a designer must engage in engineering judgement, which is a necessary part of refining and checking these models.

Having been conceived, analysed, checked, and drawn the structure, it still needs to be communicated to clients, architects, contractors, and surveyors. With approximations and sketches to hand, an engineer has to hand a means to communicate the design to those who commission an engineer in the first place.

This book documents a collection of approximate methods which have been collected over a career of over twenty five years in structural engineering design using real examples of built projects. It follows a logical progression of most common types of structure with numerous images and annotated sketches to back up the text.

Fourth Edition (2017)

Chapter 2.4 (continuous beams) has been amended for improved clarity of diagrams. For greater ease of comparison, all bending moments, shears, and reactions have been altered to a percentage of the equivalent simply supported values.

Since the book was originally published, a number of people interested in purchasing the book have asked if foundation engineering is included. In response to this, for completeness, chapter 2.13, regarding foundations, has been added to the book. This is not a comprehensive guide, but it follows the principles of the book, with a description of foundations in their various forms, and outlines some simplified calculation methods to resolve common problems. It is hoped that this will be useful to future readers. that this will be useful to future readers.

Chapter 1.1

1.1 Introduction

"Engineering problems are under-defined; there are many solutions, good, bad, and indifferent. The art is to arrive at a good solution. This is a creative activity involving imagination, intuition, and deliberate choice." *Ove Arup 20th Century Engineer*

Ove Arup, the founder of a global engineering phenomenon, the structural engineering consultancy for the iconic Sydney Opera House, could not have summed up better the art of engineering. It is how one comes up with deliberate choice, which is the essence of the art of structural engineering.

The content is aimed at structural engineering students and newly graduated structural engineers but it may also be of interest to architectural students, practising engineers, architects, and other building professionals.

In the second part of the book, specific examples with accompanying narratives demonstrate common problems and how engineering judgement might be applied to achieve efficiency of form.

The concluding chapters undertake a review of failures, which are instructive and indispensable to educating enthusiastic newcomers to the profession, together with some cautionary advice.

Complex formulae and numbers are avoided. None of the solutions is per se the correct answer. The assiduous designer should always be searching for alternatives and testing the answer for its appropriateness. In a review of structural problems that advance from the simply supported beam to more complex forms such as stadia, it is hoped that this book will assist in providing a solid introduction to the art of structural design.

There are many textbooks available that cover the areas of the theory of structures and others which will instruct in the design of specific materials. What is the need for a book which does not follow structural principles with the rigour of the aforementioned texts?

There are two categorical reasons for the provision of annotated pictorial aids to structural design, both of which involve teaching and education with respect to the subject. The first reason is to advocate and encourage a qualitative appreciation of engineering structural analysis.

The subject matter has been well documented in the work of Dr David Brohn, who has written Understanding Structural Analysis (Brohn 2005, reference 1.) which teaches structural analysis as a qualitative approach, as opposed to a purely quantitative approach. Brohn has published material and has run courses that aim to teach structural behaviour: 'as a visual language of structural behaviour that can be understood with the minimum of textual comments.'

Brohn has championed the training of young engineers to recognise structural behaviour with the use of diagrams and without reliance on numbers; a qualitative approach. It is reported that the initial tests, which involve some 20 different structural puzzles, have low scores by recently graduated engineers. Over the years, the scores have, if anything, worsened, not improved.

Professor Ian May noted on the website of the Institution of Structural Engineers regarding a study group for the teaching of structural analysis: 'The need for graduates to have a deep qualitative understanding of the behaviour of structures under load has never been greater, as practical analysis and design are increasingly based on the use of third-party software rather than hand calculations.

The second reason for this publication is to demonstrate that good engineering involves the imaginative use of structural mechanics. The intention is to encourage a visual approach to design. The work of Brohn et al. has exposed a weakness in the education of engineers.

Engineers might not have to be good at producing their own technical drawings, but at the very least, they should be able to interpret drawings and visualisations by others, architects in particular. They should be able to model in their imagination 3-dimensional representations of their design proposals. They should be able to produce hand-drawn sketches so that they may be understood by others.

The skill of conveying solutions in simple diagrammatic forms without having to rely on complex formulae will assist trainees and qualified engineers in building up their own vocabulary and store of answers to problems posed. Real examples are referenced alongside to reinforce the material.

Modern engineering practice in the construction industry likes to emphasise the importance of standards or codes. National procedures set out to ensure that calculations have a unified approach. In the United Kingdom, the profession is moving towards Eurocodes, which are intended to standardise calculation methods across the European community.

Eventually, engineers across Europe will have to master the new vocabulary where, amongst other alterations, dead loads become permanent actions, and live loads might be referred to as variable actions. The unfamiliarity of the terminology and complexity of some of the concepts serve to make intelligibility a tedious and frustrating task. No apology is made for the minimal of reference to national codes within this book.

What is structural engineering? A definition that will suffice is a design process whereby the disposition and size of elements, which constitute a framework, are defined by analysis to have sufficient strength and rigidity. Appropriate geometrical forms and sizes for the elements are chosen, which are most economical for the purpose. The design is communicated by technical drawings, with dimensional form and materials specification, to enable a constructor to build the assembly.

Structural engineering has its place in building construction, sub-sea structures, maritime structures, ships, aeroplanes, and spacecraft, to name but a few. In building construction, structural engineering involves the design of supporting structures which support, for example, floors, walls, storage tanks, mechanical and electrical services; structures that provide rigidity to prevent toppling over in the event of adverse lateral forces – wind, water, earthquake or *lack of fit* effects.

Why employ an engineer today if there is computer software available that can undertake the structural design? Engineering is the ability to assess design problems, formulate or apply an ingenious or novel solution based on experience and skill, and present this in such a way that the solution can be carried out by others.

Until computers are able to perform this skill, the world will need engineers trained to attain the aptitude and skills needed to make the engineering decisions required of the 3-dimensional computer building models of the future.

Most structural engineers tend to specialise; for instance, some work exclusively on buildings, others on aeronautical structures, and so on. Engineering principles have much in common, but engineering practice is too demanding for engineers to cover all of the fields. This book is mostly concerned with the field of architecture.

How are the budding professionals of tomorrow to be attracted from the talent pool of the schools? A glance at a syllabus for civil/structural engineering is perhaps a defining moment for the school leaver considering the course. Engineering courses offer a virtuous selection of technical subjects. A timetable might include structures, drainage, geotechnical engineering, law, management, fluid mechanics, and hydraulics; enough to wet the palate of the academically or technically inclined, but where's the joy? Appreciation of the great engineering projects of the past? The art? It is there - it simply needs amplifying, or in today's jargon, a higher profile.

Aspiring architects have a much more stimulating introduction to the subject. An extract from a website for the UCA (University of Creative Arts) has a succinct introduction: 'BA (Hons) Architecture is an ARB and RIBA validated course. The course operates as a design laboratory in which you seek radical conceptual approaches to architecture and test them through debate and through application in public space. You consider spatial practice in relation to the human conditions experienced in everyday life as well as in relation to social, political, landform, urban, and virtual environments.'

Two professions that are to form a future relationship diverge in their cultural emphasis from the beginning of their academic study. There are some good courses that bring architecture and engineering together. Bath University is a good example, which runs parallel courses for architectural and engineering undergraduates and actively encourages their collaboration and mutual enlightenment.

It is a challenge to set a curriculum in which all of the compulsory subjects set by the engineering institutions are included in sufficient depth to educate engineering students sufficiently. An engineer needs a certain proficiency in a number of subjects, including materials science, structural mechanics, foundation design, surveying, mathematics, and technical drawing. With a declining trend in applications from the young seeking more lucrative careers, attracted to the world of financiers and accountants, is it not time to address this pressing matter?

Surely, the excitement, the challenges, the intellectual demands on an engineer in the course of a career can be put across more effectively though greater emphasis on teaching both past achievements and fun in design?

Vitruvius, the celebrated Roman architect and engineer, stated that good buildings should demonstrate: 'Commodity, Firmness and Delight'. The packed syllabus of engineering courses pays high regard to Commodity and Firmness, but engineering students are seldom challenged with Delight. Is it feared that our professional visage might slip from the constraints of impartiality? Why do engineers not appear to be encouraged to delight in the process of fitting the dry mathematical concepts of engineering into the design process? Why not discuss massing, the language of detail, or even its masculinity or femininity?

Many engineering undergraduates may never touch on architecture in their courses, but architectural students are mostly committed to studying structural design. A brief survey is sufficient to confirm that many architects are apprehensive about the study of structures and keep a distance from its practice during their careers. It is common sense to provide guidelines for a professional to whom the planning and conception of a building can only work if it fits within the constraints of the supporting structural skeleton. Often, for simple buildings, a structural engineer is not involved. Architects who spend a career in smaller scale projects generally gain a good intuitive understanding of structures, but it is more difficult for architects involved in larger scale projects.

A selection of design guides for architects, based on simplified structural principles, is available for the teaching of structural design to architectural undergraduates. It is hoped that this book might also provide a reference for architects, giving insight from the viewpoint of a practising engineer in the world of architecture. It may even provide possible structural solutions with which a practising architect may then be able to prompt the project structural engineer to try an alternative approach.

Apprenticeships have been largely replaced today by graduate training after the university education is complete, which entails a good deal of time to connect theory to practice, a frustrating journey that seems to the novice to go backwards before moving forwards. Enthusiastic to enter the real world, the graduate has to learn to apply theory, often forgotten, only to have to satisfy the demands of the building design team, all their own agenda: clients, architects, builders, project managers, and quantity surveyors, to name but a few.

However much an engineer's education is filled with complex analysis; however strong the emphasis is on codes of practice, the professional engineer, just like the medieval journeyman, has to solve real problems. Those who must conform to the architectural form of buildings that have to be built by contractors grappling with the perplexing task of putting a building together.

Engineers have to work in sympathy with builders, who have to organise the labour, build the temporary support structures and put together large structures safely, which need mechanical lifting devices to raise heavy sub-assemblies.

Engineering is enjoyable. It can also be energy-sapping, puzzling, frustrating, and sometimes tedious in achieving its aims. But it is an engaging and challenging profession. This book provides a link between structural theory and structural design, guiding undergraduates struggling with their project assignments through to graduates employed on real projects.

Nomenclature and units in the book

Dead loads are used to describe the self-weight and service loading combined. Live Loads are used to describe the applied loading of the occupancy.

All units are SI units:

kN – forces;
kNm – moments and torsion;
m – spans and overall dimensions;
mm – deflections

Chapter 2.1

2.1 Efficacy Balance and Grace

"If you have built castles in the air, your work need not be lost; that is where they should be. Now put the foundations under them." *Henry David Thoreau*

Efficacy, Balance and Grace

The three pillars of good architecture espoused by the Roman architect Vitruvius are: 'Commodity, Firmness and Delight'. It is instructive to reinterpret the message to suit engineering. Maybe *efficacy, balance and grace* also fit the picture.

Efficacy may be defined as the power to produce effects, which is not just efficiency but the ability of a given design to perform its intended function. It is a word which embraces economy in terms of minimising materials used. Well-designed kitchen utensils often seem to be beguilingly simple but with a fitness for purpose, which is a pleasure for the user. Take a knife, for instance, which has a handle fitting snugly in the hand and a blade which has enough width to cut through tough vegetables. It also possesses balance, which makes it easy to hold.

Balance is the means of achieving equilibrium or stability. A bridge with a finely honed weight distribution, such as the historic and well-known Forth Railway Bridge is a beautifully poised structure. In some circumstances, the structural elements are expressed in the external envelope of the building. Part of the art of sculpting the form of a building is to arrange the elements such that they achieve a balanced whole.

Grace is the way in which the sculptural form might transcend its mere function and can become an object of real beauty and delight. Architects are taught to appreciate grace. Engineers should engage in this vitally important component of design. At the least, it should be appreciated that achieving grace of structural form is a skill which takes training and education.

Unlike fine art, architecture and engineering are bound by the necessity of producing buildings and bridges which are built to safety standards and which are necessarily utilitarian.

A good designer will have in mind public perception of their designs: is the structure to look solid conservative and safe, or is a more dynamic form demanded? It might be a monument on a traffic island in a busy highway. Clarity of form is desirable.

A current architectural trend, which follows from the postmodernist design world of the 1980s and beyond, has been to play visual tricks with the external form of structures. The balance of structure may be made to look decidedly precarious; visually setting up a conflict within the mind of the viewer; a struggle to accept the reality of a structure which looks as if it might fall over but self-evidently does not.

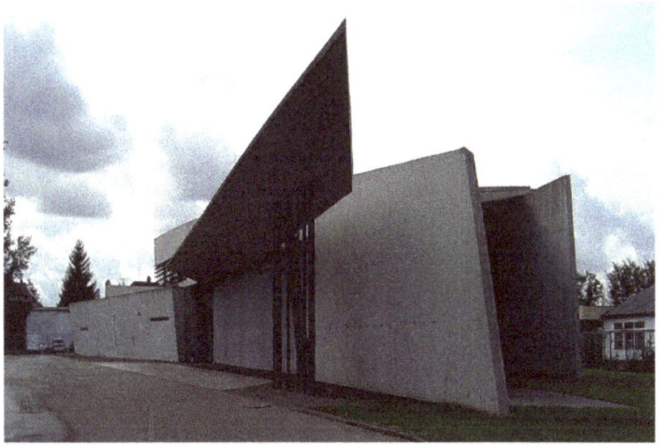

Figure 2.1.1 Vitra Fire Station

The late Zaha Hadid, a contemporary international architect, was such a visual games player. One of her first built designs was the Vitra Fire Station in Germany (**figure 2.1.1**). It looks nothing like the archetypal image of a fire station with a pointed canopy standing on leaning stilts which juts out from the main building, a fluid form of concrete leaning walls. The building appears to have little sense of structural logic on one hand, but on the other, it does have a distinguishable form.

Understanding and Communicating Structures

An understanding of structures is helped by observing more difficult buildings like the Vitra Fire Station. Studying the work of contemporary Spanish architect and engineer, Sanitiago Calatrava, is a lesson in the art of how a perception of structures and the development of elegant structural design is aided by techniques more familiar to sculptors than engineers.

Figure 2.1.2 is a sketch of a sculpture by Calatrava, which is the representation of a leaning torso - a human backbone in the act of leaning forward. Five cubes, which represent the vertebrae as compression elements, have a centre of gravity located in the middle cube, located away from the supporting conical post. The cubes are interlinked with short compression elements.

To stop the whole construction falling over, a wire is tied down to the ground, like the *tendons and muscles in a spine*. Calatrava demonstrates lucidly, based on a natural form, the interplay of components in a structure. The cubes, as compression elements, are solid and stumpy.

The tension wires are thin. In the natural world, elements in compression tend to be stouter than those in tension owing to the phenomenon of buckling: the curvature of a slender shape under the action of compressing forces.

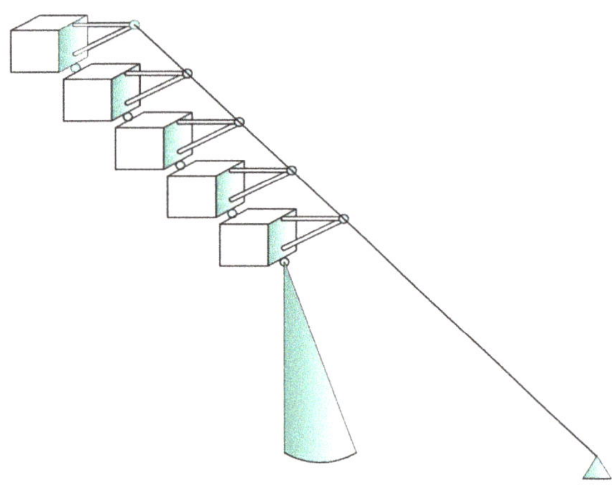

Figure 2.1.2 Calatrava Backbone Sculpture

A suspension bridge has a beautiful natural form. (See **figure 2.7.2** in **chapter 2.7**). The compression elements - the supporting towers or piers and the supported deck - are stouter in form than the tensile cables.

Furthermore, the fluid curvature of the cables mimics the bending moments (flexure) in the system, which is a function of the span.

A famous historical demonstration which was shown to the public of the Forth Railway Bridge (**chapter 2.6, figure 2.6.12**) is shown in **figure 2.1.3**.

The central span of the bridge is carried by two outer balanced cantilever piers. The three men in the picture are a vivid representation of the bridge. The arms stretched out illustrate the tension elements. The struts or compression elements are straight and solid-looking rods, and the pallets of bricks show how the ends of the supporting cantilevers are tied down.

At the time of building the Forth Railway crossing at Queensferry in Edinburgh there was public concern shortly following a disastrous bridge failure over the river Tay in 1879. An engineering wonder of the time, multi-span iron girders raised on a series of iron towers was destroyed during a winter storm, taking a train with passengers into the river.

The Forth Bridge was completed in 1890. It brought about the commission of two English engineers, Sir John Fowler and Sir Benjamin Baker, who replaced John Bouch, the unfortunate designer of the Tay Bridge.

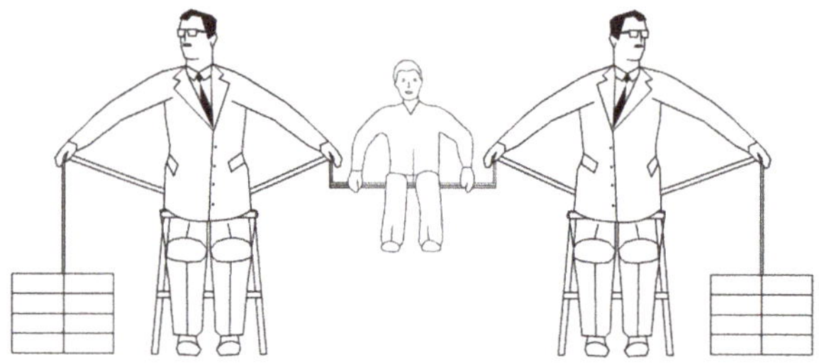

Figure 2.1.3 Forth Bridge Demonstration

It was Britain's first major construction fully in steel. Its design was somewhat different from the Tay Bridge. Much of the reason for choosing the design was that it looked stable and solid, in great part a response to public concern over the Tay Bridge disaster.

Engineers are employed to produce structures which are not only fit for purpose but are also perceived to be thus.

Linking Past and Future

Peter Rice of Ove Arup Partnership collaborated with architect Martorell, Bohigas, Mackay (MBM), together with Jaume Freixa and Jose Maria Baquero in the Seville exposition in Spain in 1992. Peter's contribution, with architects MBM, was known as the Pavillion of the Future. See **figure 2.1.4.**

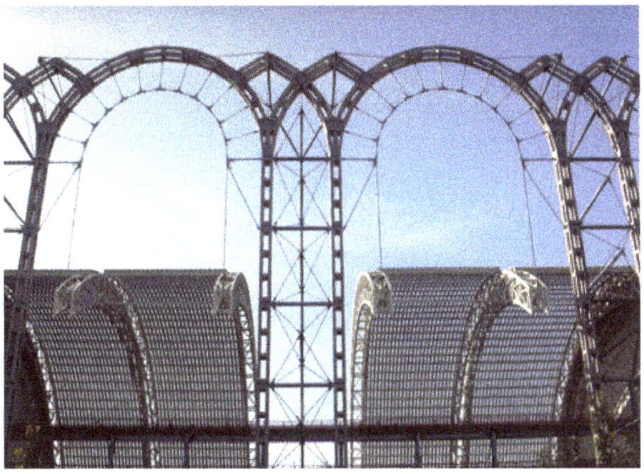

Figure 2.1.4 Peter Rice – Pavillion of the Future – Courtesy of Arup

The structure is a free-standing framework representing the past and future, from the historic arches of the Segovia Aqueduct (see **figure 2.1.5**) to the futuristic lightweight arches balancing tensile and compressive components. Pointed arches formed with columns comprising steel ties and rods, which are lightweight in form. Its rhythm is achieved by the repetition of the arches.

The fundamental difference in the analysis was that the structural forces within the frame could only be analysed by a reiterative, non-linear assessment of the geometrical deformations of the structural elements.

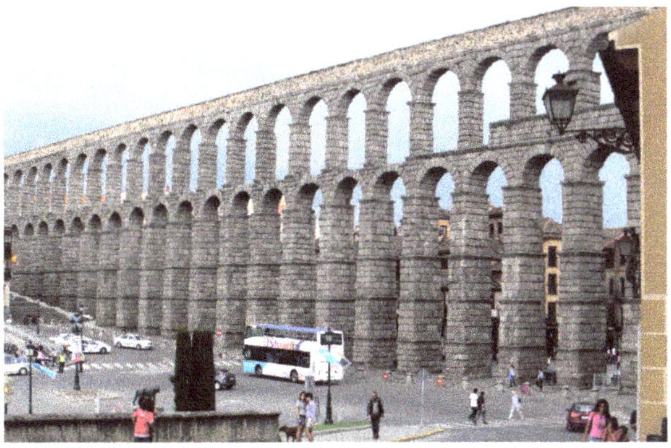

Figure 2.1.5 Sergovia Roman Aquaduct, Spain

This is distinct from the everyday linear structural analysis that most engineers use. It was necessary because the actual geometrical deformations of the structural parts are as significant as the material deformations.

Like Calatrava's sculpture in **figure 2.1.2**, it has clarity of expression of structural parts: the solid granite compression elements and the relatively thin steel ties holding the assembly together.

Structures are static forms bound by gravity and the vicissitudes of the physical environment. Paying homage to the limitations set by design is the route to successful aesthetic forms.

Calatrava's leaning torso sculpture, the Forth Railway Bridge and the Seville arches are good illustrations of the simplicity and seeming effortlessness of good design. Each component part obeys a structural logic, combining to form a graceful structure, achieved by careful consideration of balance and structural efficacy.

These examples have been selected for the fact that, whatever the complexity of the structure and selected method of analysis, the designer must justify aesthetically and functionally the size of the components, the materials and how they are connected together, all of which are interdependent.

To a great extent, the initial design process is intuitive. It involves early discussions with architects and other team members. Notwithstanding the brilliance perhaps of the original concept, a design has to be tested and its suitability reviewed. Rather than turn immediately to a computer, a good conceptual designer is able to make thumbnail sketches and to undertake crude calculations.

The following chapters (**2.2-2.12**) will show how short-cuts and approximations may be used to design anything from stadia roofs to checking masonry bridges. Approximate solutions give a visual aid furthermore to explaining structural action; much appreciated by clients.

Efficacy, balance and grace is available with the minimum of fuss and complex computer models helping the graduate or student to attain true engineering judgement.

Chapter 2.1

Brief Summary

An engineer who is particularly alert, with patience and inquisitiveness, might, by serendipity, discover simplified approximations and useful analytical shortcuts when checking their computer models. It is contended that there is a gap in educational material which connects conceptual design with approximate methods. The inexperienced designer is at risk in their early career because their analytical models are not a true representation of reality.

In the book 'Everything is Negotiable' (Kennedy 1997, p.76, reference 2.), with respect to contracts, an interesting quote is worth noting: *'In any 12 month period over 400 claims for faulty design, inaccurate calculation and inadequate checking of work are made against consultant engineers by their clients from all corners of the globe'*. This quote is some 30 years old. Of course, countless structures are built without problems; structural engineers are mostly responsible and successful in designing safe structures.

What can be done to ameliorate the problem of unrealistic computer modelling, which might lead to faulty design? One positive step is to address at universities and in the design office methods of appraising designs which cut through the complexities and ask a design *'what if?'* questions, which are essential to attaining an understanding of the characteristics pertinent to any given structure.

The London Wembley stadium roof is suspended from an arch. This arch may look like a complicated looking lattice arrangement of steel elements, but the forces within the arch components all conform to the overall static equilibrium. The diagonal arrangement of the cables from which the roof is suspended helps in restraining the arch from buckling. Stephen Morley came up with this concept: a brilliant engineer who has designed many dazzling stadium roofs, including the catenary roof structure over Chelsea north and South Stand and the Sydney 2000 Olympic Stadium.

Morley bases his concept designs on sketches and simplified formulae. The concept for the saddleshaped roof of Sydney Stadium was inspired by the shape of a potato crisp. In **chapter 2.8** a conceptual approach to the design of another Stephen Morley project – the Golpharm Stadium in Huddersfield – is followed using the simple methods gained in the preceding chapters. With such tools, the most intricate or daring structures may be conceptually designed.

Demonstrating to the design initiative that it is possible to use simple formulae to conceptualise grandiose structures can serve only to encourage and inspire. The aim is not to denigrate the hard-won achievement of learnt structural theory. On the contrary, it reinforces this education and reduces the risks of misunderstanding computer models. Furthermore, simplifications advertise to the general public, to whom strong images and simply explained concepts are more easily digested, the challenging art of structural design.

Some familiarity with structural design and analysis will be necessary for effective use of the examples. This should not deter the interested architect who is looking for a better intuitive understanding of how good structural designs are configured. It also serves as a reference document for some of the more efficient forms of structure which engineers have to handle. The analysis, which precedes the design, is not resolved with accurate figures in one go. This is the what if? stage. *'What if I make that joint a pin rather than fixed?' 'What if I move that hinge?'*

Many of the best engineers and architects make observations, and build up a repertoire of designs. Observing buildings and structures is instructive in a way that no university lecture can provide. Look at buildings and pose questions. How does the first floor cantilever of that tower block stand up? Why is it there? Why does it seem to be too thin? Our eyes are indispensable tools.

Finally a repertoire of solutions and approximate methods is most useful. It may take time to acquire but its attainment is the key to a more purposeful and enjoyable approach to structural design.

Chapter 2.2

2.2 Simply Supported Beams

Every student of structural engineering is introduced to the most basic but perhaps most useful of components: the simply supported beam. It starts with a reintroduction of the basic bending, shear and deflections. It is followed by an anecdote by Galileo to show how easily intuition can fail us; going back to the simply supported can lead to safer assumptions of behaviour. Finally, some useful approximations are recorded of benefit to the engineer engaged in a more sophisticated analysis.

Before the advent of iron, then steel and reinforced concrete, the principle building materials for walls were brick, stone or some other materials such as timber stud-work. For floors and roofs, although stone lintels could be used in limited circumstances, timber was the main structural material used; rafters or joists as the main load-carrying elements, and planks to form a floor or roof deck.

The specific property which meant that timber was suited to floors is that it is essentially a ductile material; timber fibres can resist tensile strains. When a log is placed over a stream, between two supporting rocks, it can carry the weight of a person or animal in flexure (bending) as a ductile material.

It took scientists and engineers some 250 years to advance the science of structural engineering to a point where there is now reasonable confidence that the behaviour of simply supported beams may be predicted. This chapter is appropriate for relatively shallow beams which behave in flexure.

Where the engineer has to take care is when beams are deep in relation to their spans. A deep beam may loosely be described as one in which the ratio of its depth to span is 3 or greater. Deep beams essentially work in shear rather than flexure. **Figure 2.2.6** gives a reasonable indication of when beam behaviour deviates from the purely flexural.

The famous monument of Stonehenge has stone beams atop stone pillars. One might ask: are these not acting in a flexural manner? Correct, they are stone lintels or beams. However, they are relatively deep in relation to their span, which allows them to make use of this bulkiness by carrying their own weight in two possible modes. **Chapters 2.8** and **2.10** elaborate further on these principles.

In the next section, the basic elements of beam analysis will be reviewed: *shear forces, bending moments and deflections*.

Shear Forces

Figure 2.2.1 shows a basic set of reactions for three cases of loading: uniformly distributed loading (UDL) and point loading (concentrated loading from another beam/person/post etc.). An important property of a simply supported beam is that it is statically determinate. The reactions may always be found by resolving two equations. Although simply supported beams, as will be seen in **chapters 2.3** and **2.4**, are not necessarily the most structurally efficient or architecturally desirable way of spanning across a space, the determinate property can be very useful in its predictability.

The reactions are concerned with the shear in the beam. Checks may be made to ensure that the beam has sufficient shear capacity to carry load and that the beam support reactions may be found. For symmetrical loading, the support (or end) shear is half the total applied loading.

Deflection (or Deformation)

An important criterion to assess for a beam is the stiffness of rigidity. The rigidity of a structural system is generally concerned with an area known to engineers as the serviceability of a structure.

The two most commonly used deflection equations are shown in **figure 2.2.3.** These are generally recalled by experienced engineers and are very useful to utilise for other loading conditions, particularly when an approximate design solution is being checked for its validity.

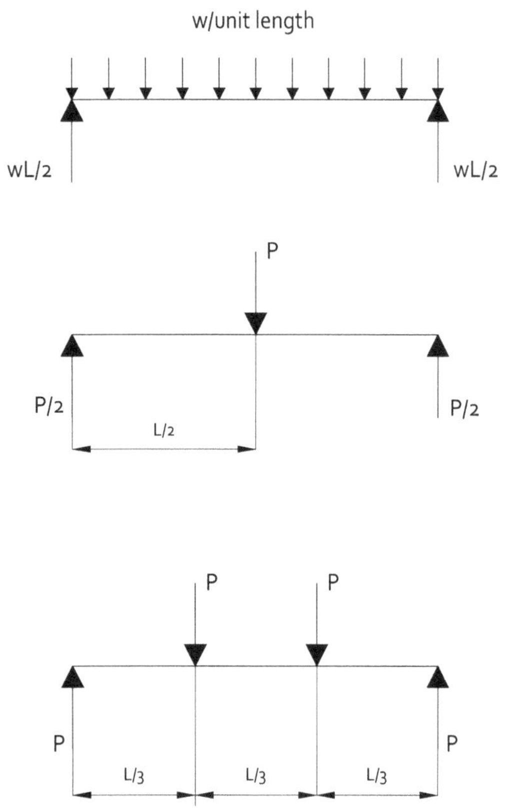

2.2.1 Simply Supported – Shear Forces

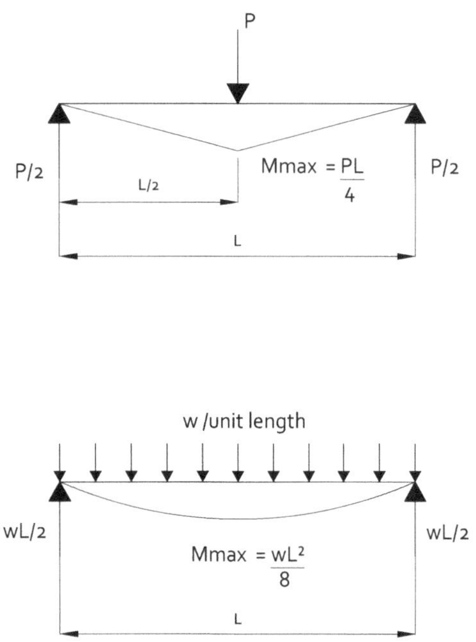

2.2.2 Simply supported – Bending Moments

Bending Moments

This phenomenon is the curvature of a beam under applied loads which are perpendicular to its span, inducing tensile and compressive strains in sections which are equal in magnitude.

The central and maximum bending moment in a simply supported beam with a Universally Distributed Loading (UDL) is the engineer's most commonly used formula: **WL2/8**. See **figure 2.2.2**.

The maximum bending is in the middle or centre of the beam. It is simple and predictable. A beam thus sized does not need any special supports and will not depend on the builder, particularly understanding the intentions of the engineer.

Built-in or *encastré* supports are discussed later. Refer to **figure 2.5.1.**

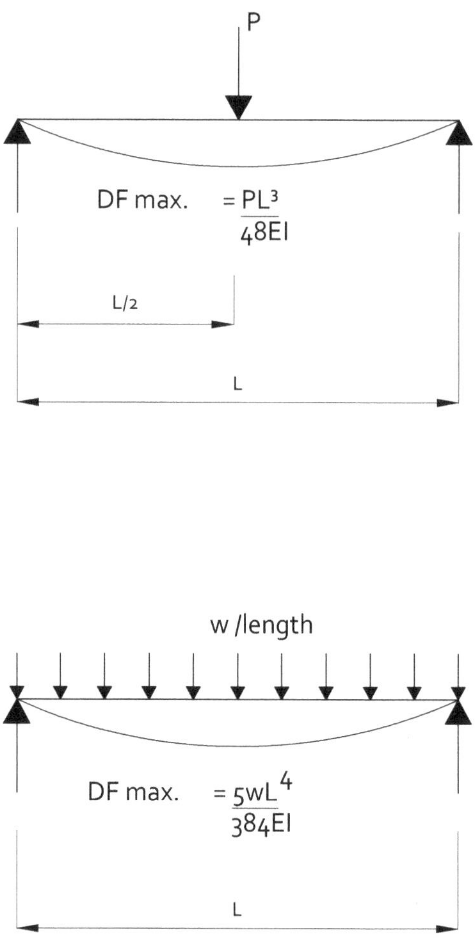

Figure 2.2.3 Simply Supported Deflections

Advantages of Simple Supports

Galileo, in his two new sciences, discussed the problem concerning the storage of marble columns, which were mounted on blocks to keep them from gathering dirt: *'A large marble column was laid out so that its two ends rested on a piece of beam; a little later it occurred to the mechanic that, in order to be doubly sure of it is not breaking in the middle by its own weight, it would be wise to lay a third support midway; this seemed to all an excellent idea.'*

It transpired that this was not such an excellent idea. Far from reducing the breakage of stored columns, it actually resulted in a number of columns breaking their backs over the central support. The simply supported column had now become a statically indeterminate *continuous* system.

Figure 2.2.4 4 illustrates the problem. The weight of the column (represented as 'W', which is a uniformly distributed load per unit length of beam) has a flexural moment expressed as WL/8.

In **chapters 2.3** and **2.4** cantilever and continuous beams are discussed, but for now, the problem of the column is that any *give or springiness* in the outer supports will cause an increase in the weight borne on the central support.

This redistribution of loading increases the bending moment in the centre of a column, which for solid supports approximates to **WL/8**. For increasingly *spongy or soft* supports, the bending moment can reach **WL/4**; twice the original simply supported value. The column supported thus is now highly sensitive to any give in the supports. It is no wonder that there was an observed increase in the breakage of the columns.

Perhaps the builders of Stonehenge had a better structural appreciation than previously thought **Figure 2.2.5** shows the stone lintels fortuitously split over their supporting posts - **WL/8** prevails?

In **figure 2.4.13** an example demonstrates how breaking up a continuous ground beam into simply supported elements, whilst not as structurally efficient as a continuous beam over multiple supports, has great benefit in that pile loads (beam reactions) may be easily predicted.

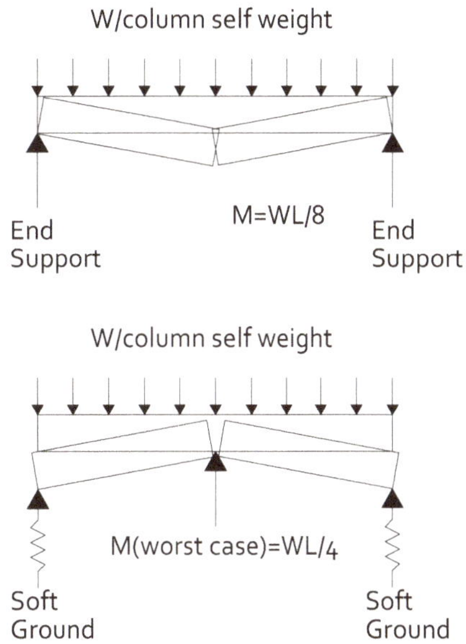

Figure 2.2.4 Galileo's Column Storage Problem

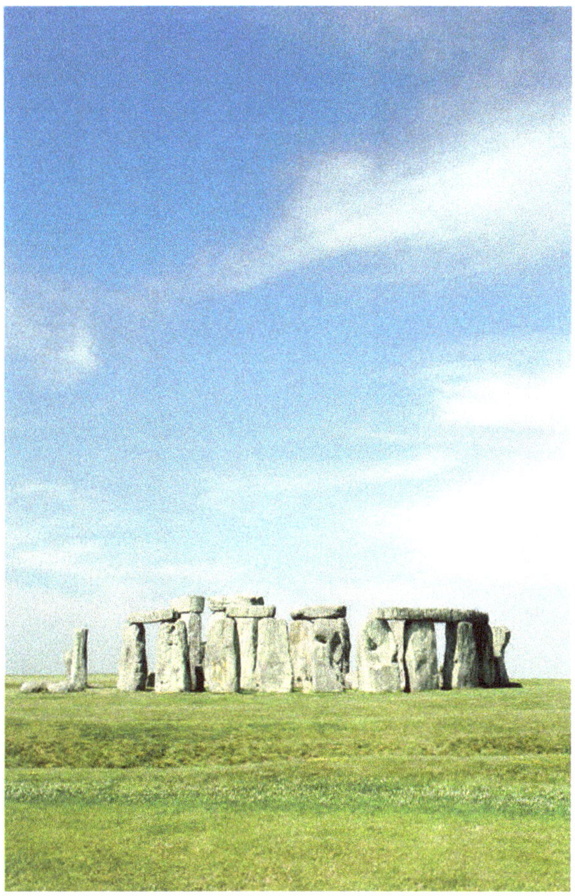

Figure 2.2.5 Stonehenge Lintels

Aside

Beware Shear Deflections and Deep Beams

Deep beams are used or occur in buildings, for instance, where a wall spans across a space and carries its own weight without a supporting beam. Deep beams (where the ratio of depth to span is 1/3 or greater) might need two deflection calculations: deflection based on its flexure or bending and deflection based on its shear. Deflections discussed in this book mostly concentrate on bending deflections. Since most beams are not selected as deep, they have a specific use: when analysing concrete lift shafts or on the façade of a building, where the full floor-to-floor depth is used structurally. See figure 2.2.6.

Chapter 2.11 examines deep beams in more detail.

Chapter 2.6 *describes the effect of shear deflections in a truss.*

In figure 2.2.6 *three beams are shown together. Taking a homogeneous, ductile material-like steel, the relative contribution of the 'shear' deformations to the 'bending' deformations is compared in table 2.2.1, where D is the depth, and L is the span. For the three beams shown table 2.2.1 summarises contribution of shear to the overall deflection.*

Shear deformations may be ignored for all practical purposes for any beams up to a depth-to-span ratio of 1/3.

Ratio (Depth/Span) D/L	Contribution of Shear Deformation %
1/8	1.5%
1/3	10%
1	90%

Table 2.2.1 Deep Beams Span Depth Ratio

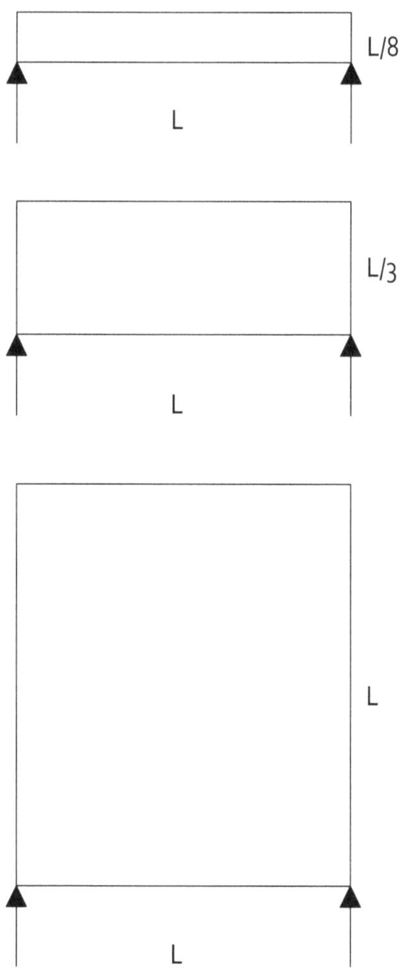

Figure 2.2.6 Deep Beams Compared ref. Table 2.2.1

Some Useful approximations: Simplification of Loading

Many structural flooring systems are a combination of *universally distributed loads* (UDLs) and point loads. Joists carry floor decks (concrete or timber deck) as a UDL, which are then supported off girders, imparting a point load onto the girders. If a girder sits diagonally across a rectangular floor plate, then the loading distribution may be triangular in shape.

This adds complexity to the calculation, which may be readily transferred to a computer model. How may an engineer make approximate assessments of the computer output?

Convert Complex Loading into Simple UDL or Point Load

In the examples which follow, different loading conditions are examined. Since the forces and deflections are readily checked for a beam with uniform UDL or central point load, a number of examples of conversions of more complex loading patterns into an *equivalent UDL* or *central point load* are compared with the substituted loading condition:

Figure 2.2.7: symmetrical triangular loading is replaced with a UDL equivalent to **2/3** of the peak intensity of load.

Figure 2.2.8: triangular loading with the peak occurring on one side is replaced with a UDL equivalent to **2/3** of the peak intensity of load.

Figure 2.2.9: third span (twin) point loads (P) are replaced by an equivalent UDL equal to **3P/L** or 3xload/span.

Figure 2.2.10: Third-span point loads (P) are replaced by an equivalent central point load equal to the sum of the **third-span loads** (P).

Figure 2.2.11: Triple point loads (P) equidistant are replaced by a UDL equal to **4P/L.**

Figure 2.2.12: Four point loads (P) equidistant are replaced by a UDL equal to **5P/L**.

Figure 2.2.13: A demonstration that loading within the *middle third* of a beam may be assumed to act in the midspan of a simply supported beam with an accuracy of 85%.

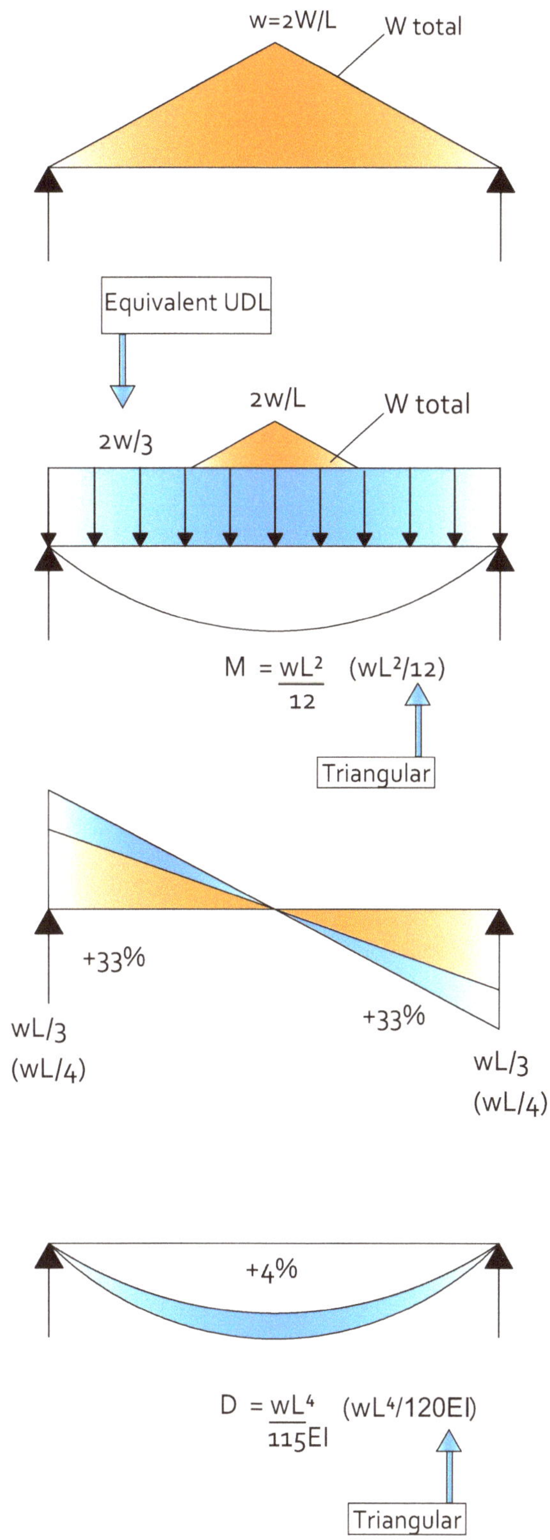

Figure 2.2.7 Symmetrical Triangular Loading

Triangular Loading (Equivalent UDL)

Triangular loading equations are not readily called to hand. However, safe approximations are possible by converting the load to an equivalent UDL. If a girder spans diagonally across a rectangular building or room, the loading is not a UDL but triangular. w/unit length is the peak load.

Formulae for two triangular loading cases are compared with an equivalent UDL, which may be used as an approximate substitution. See **figures 2.2.7** and **2.2.8**.

If 2/3 (two-thirds) of the peak loading is assumed to act as a UDL, the formulae for the simply supported beam in **figure 2.2.2** and **figure 2.2.3** may be used.

The formulae for bending, shear and deflection are shown in **figures 2.2.7** and **2.2.8**.

Where the triangular loading differs from the equivalent UDL, the formula is shown alongside in parentheses: e.g. (WL/6).

For example, in **figure 2.2.8** the equivalent UDL bending moment is **WL2/12**, and the triangular bending moment is **wL2/16** (33% overestimate).

In conclusion, the equivalent UDL overestimates or equals the triangular load bending moments and deflections. This is a useful shortcut when modelling or estimating internal forces and deflections.

Take Care with Shear

Shear is obtained simply from triangular loading (the formulae are in parentheses in **figures 2.2.7** and **2.2.8**).

These should be used rather than the equivalent UDL since they exceed the equivalent UDL values.

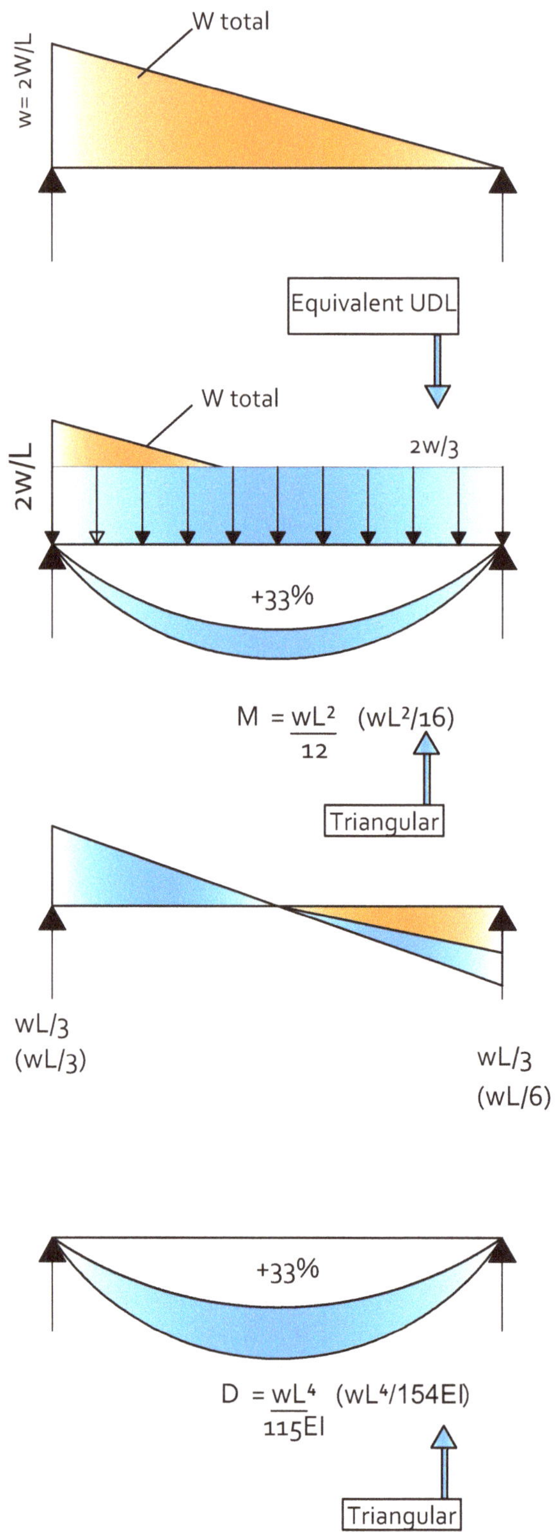

Figure 2.2.8 Asymmetrical Triangular Loading

Point Loads –Approximate Methods

The remainder of the chapter reviews a number of examples of point loads by taking the familiar UDL or simple point load acting midspan and looking at how the beam formulae compare for variations: when the loading is split into two or more point loads (acting at equidistant locations) in **figures 2.2.9** to **2.2.12** or when a single point load is moved along a beam from near the support to the midspan (**figure 2.2.13**).

At university, the student of engineering is taught exactitude: *a load of 2kN acts 2m from the support of a 10m span beam*, for instance. A quick check shortcut is useful to validate computer analysis.

The question for the engineer at the computer is: how can output be quickly and easily verified? The exact way is to check the input carefully. However, this does not give the engineer any indication of what should be expected, nor does it give any short-cut or quick verification.

The reason behind these concluding examples is to demonstrate that smoothing out the variable loading and repackaging as a familiar and quickly solved approximation, such as a simply supported beam with UDL, can yield answers often surprisingly close to the point loading case.

In **figure 2.2.9** and **2.2.10**, the substituted loading case is shown in parentheses: for instance **figure 2.2.9**, bending moment midspan is (**PL/3**).

Twin Point Loads Substituted (Equivalent UDL and Central Point Load)

In real buildings, a typical joist arrangement, which is supported from say, a roof beam, demands of the engineer a value judgement. For instance, a short-span trimming beam may be needed where a new roof-light is to be formed. The new trimmer may have two or three joists, which are to be cut and re-supported on the new beam. For a quick estimate, it is arduous to work out for each rafter the end reactions applied to the proposed trimmer and then to look up the formulae.

For two point loads, two methods are presented in **figure 2.2.9** and **figure 2.2.10**: respectively, an *equivalent UDL or a central point load.*

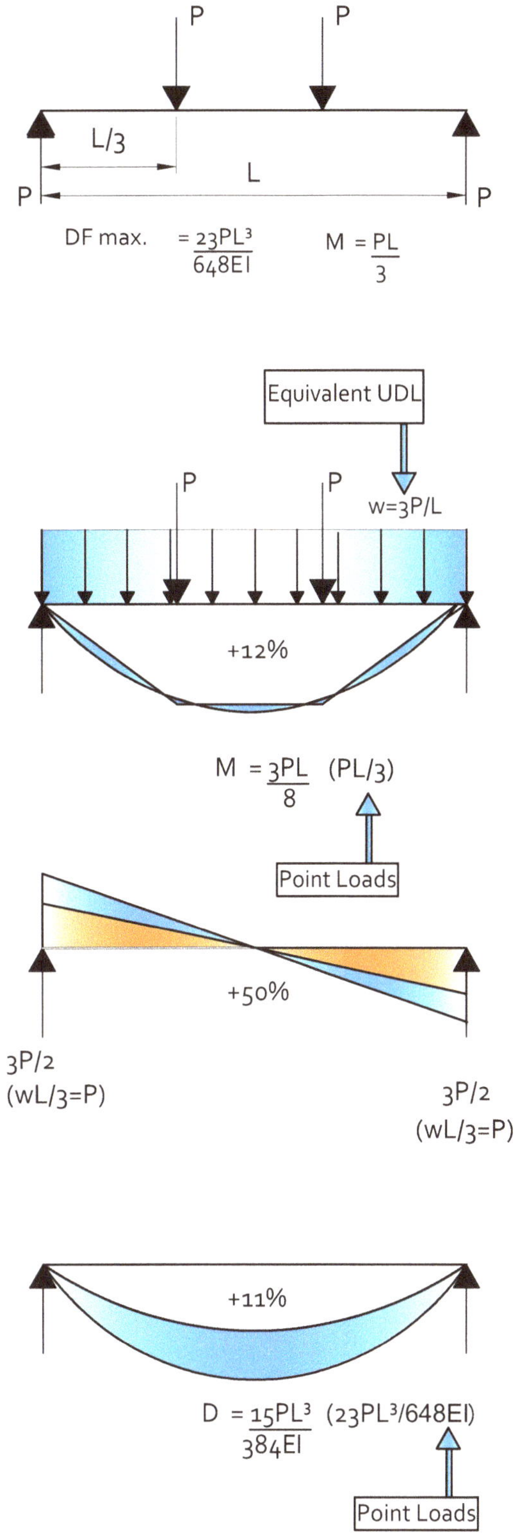

Figure 2.2.9 Twin Point Loading Equivalent UDL

Equivalent UDL

Figure 2.2.9 shows the formulae for a beam with 2 equidistant point loads. To obtain an equivalent UDL, **W=3P/L** (in other words, the point load divided by its spacing).

The formulae are summarised in **figure 2.2.9.**

In all counts, the equivalent UDL method gives greater values than the point load but not by a large margin for the bending and deflection. The percentage differences are shown in **figure 2.2.9**.

Equations for the substituted loads are shown in parentheses.

Central Point Load (Doubled)

Figure 2.2.10 shows the formulae (as in **figure 2.2.9**) for a beam with point loads. To obtain an equivalent central point load, sum the point loads (which may sometimes differ) and assume they act as one point load in the midspan position.

The formulae are summarised in **figure 2.2.10**.
The central point load yields a 50% greater bending moment, but the shear is the same (only true for exactly the same third-span point loads). Also, the deflection is close to 17% greater for the central point load approximation. Familiarity with this method allows for a reduction in the bending moment when using the approximation.

Appropriate Design

It can be seen that the central point load substitution in **figure 2.2.10** is not as accurate in terms of the bending moment examined in **figure 2.2.11**.

Sometimes, an engineer is called upon to check the existing lintel over a door, for instance, where there is an arrangement of point loads from posts over. Often, the lintel is found to have a flexural capacity well in excess of the loading. The quick method is most useful to avoid over-elaboration.

In terms of a similar exercise for shear checks, the engineer can assume the point load is next to the support and check for worst-case shear to the full load.

Should the beam fail this test, a more refined check can be undertaken. Rational engineering is often a matter of working from simplistic (*safe*) checks to a more refined solution if needed.

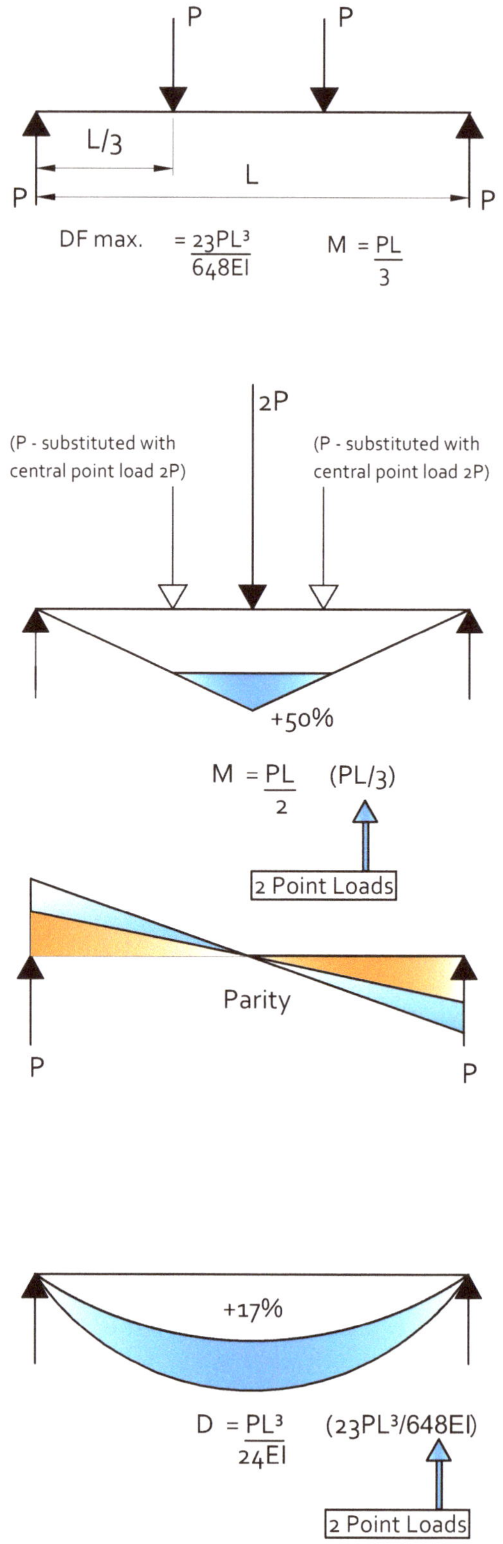

Figure 2.2.10 Twin Point Loads
Equivalent Central (double) Point Load

Triple Point Loads ....and Further Subdivisions

From the above, it is clear that for a progressively greater number of subdivisions with equal point loads, for instance, where 4 joists or more sit on a trimmer, the equivalent UDL method will yield more and more accurate results. Simply divide the calculated point load by the spacing between (i.e. w=**5P/L** for **4** point loads see **figures 2.2.11** and **2.2.12**).

This form of estimation can simplify the concept calculations and reduce the risk of error. The more iterations or sub-calculations needed to obtain the solution, the more there is a risk of an arithmetic error.

There is no reason why they cannot be used for a final design if the beam weight is not critical.

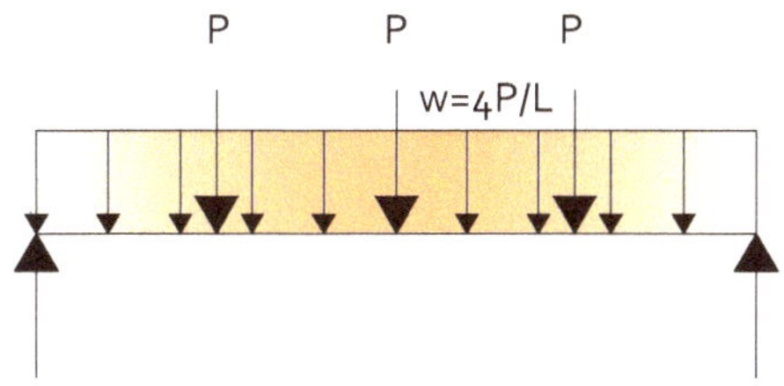

Figure 2.2.11 – 3 Point Load Equivalent UDL

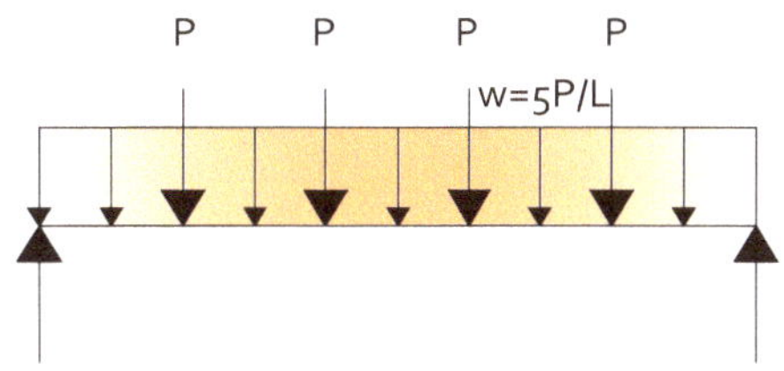

Figure 2.2.12 – 4 Point Load Equivalent UDL

Single Point Loading – Deviation from the Central Point Load

All experienced engineers are familiar with the central point load: Maximum bending moment (PL/4), maximum shear is at support (P/2) and maximum deflection ($PL^3/48EI$).

Point Load within the Middle Third

Figure 2.2.13 investigates the numerical comparison of a simply supported beam with central point load with cases where the load is moved towards the left-hand support.

Just how far does the point loading have to deviate from the centre before the above equations become inaccurate?

Ref. **figure 2.2.13** consider the designer assuming that a load at the *middle third point acts midspan*; this is the *second from top example*.

- Maximum bending moment 12% overestimate
- *Shear 60% ...Note: calculation of support shears should assume actual position if it is to be checked.*
- Deflection 15% overestimate

It may be concluded from **figure 2.2.13**, that as long as any point load is within 0.35L from the centre span, then the bending moment and deflection may be estimated on the assumption that it is in the centre of the beam. *Shear should be checked for the actual load position.*

If the point load is near the support, the deflection and moment are significantly less (P at 0.1L).

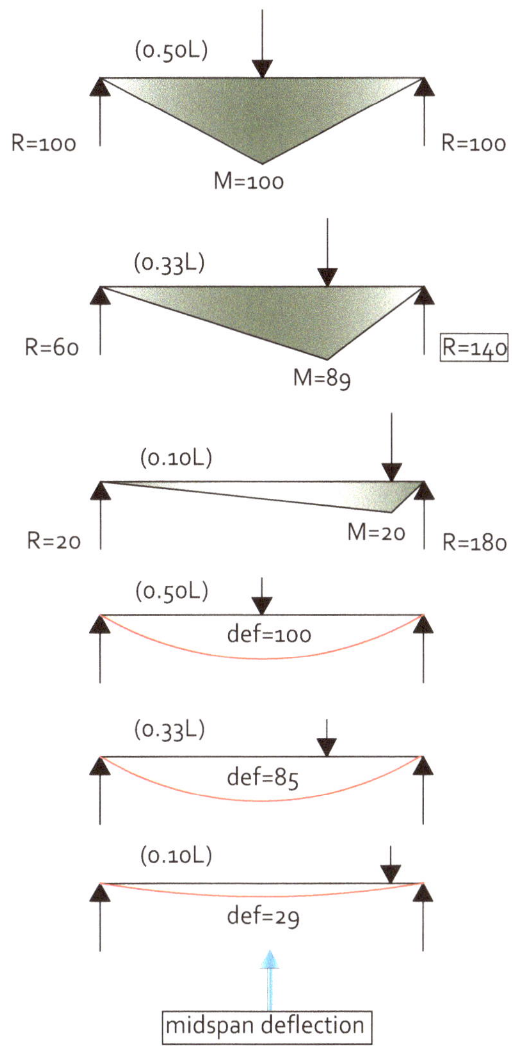

Figure 2.2.13 – Point Load Varying Position

Chapter 2.2

Quick Note Summary

1. Assuming simple support is a cautious (or conservative approach) if the real support condition is not known.

2. Uniformly distributed loading (UDL) on a simply supported beam: Bending moment (wl), shear (wl/2) and deflection (5wl4/384EI)

3. Central Point Load on a simply supported beam: Bending moment (Pl/4), shear (P/2) and deflection (Pl3/48EI)

4. Various estimates may be applied using the above, which are generally remembered by all engineers.

5. Beware the continuous sinking support (see Galileo's problem on the stored column) – moments and shears may be altered in a detrimental manner simply by adding another support (unless all supports are solid).

6. Calculating the deflection of a deep beam using bending theory causes an underestimate of the overall deflection – in addition, shear deformation should be considered. A deep beam may be defined as a beam having a depth/span ratio greater than 1/3.

7. Any triangular loading may be estimated by substituting an equivalent UDL of 2/3 (or 67%) of the peak load.

8. For two point loads within the middle third of a beam of equal value, the bending and deflection may be estimated by substituting a single central point load of twice the value of the point loads and using the formulae in 3 above. Shear is relatively easy (equal to the point load).

9. Point loads in pairs in the third span positions (of roughly equal value) or nearer the centre span may be substituted by a UDL (w=3P/L) or central point load of their combined value. Note the bending moment may be overestimated by up to 50%.

10. If a single point load is within the middle third of the beam, if it is assumed that the load acts in the centre of the span, this will overestimate the bending moment and deflection. Shear should be separately calculated.

11. For a UDL loaded beam, the deflections within the middle third of the beams are within 80% of the maximum. The deflection at 0.1L (10% of the span) from the support is 30% of the maximum midspan value.

Chapter 2.3

2.3 Cantilevers

Cantilevers are a useful and often dramatic expression of structural form. In simple terms, they are fixed at the support so as to limit or eliminate rotation and free at the end. The fixed end exhibits greater flexure and the free end greater deformation when compared to the mid-span point of an equivalent span simply supported beam. The basic principles are examined, followed by an exploration of the effects of the addition of a back-span. The concluding part shows how cantilevers may be used in real structures, with methods of achieving architectural expression.

Galileo made an early attempt to understand the mechanics of a beam built into a wall. His celebrated diagram of a cantilever is illustrated in **figure 2.3.1**.

The equations in **figure 2.3.2** for the bending, shear and deflection took some centuries to resolve after Galileo's attempt to try to predict the breaking strength of a wooden beam, which was unfortunately wide of the mark.

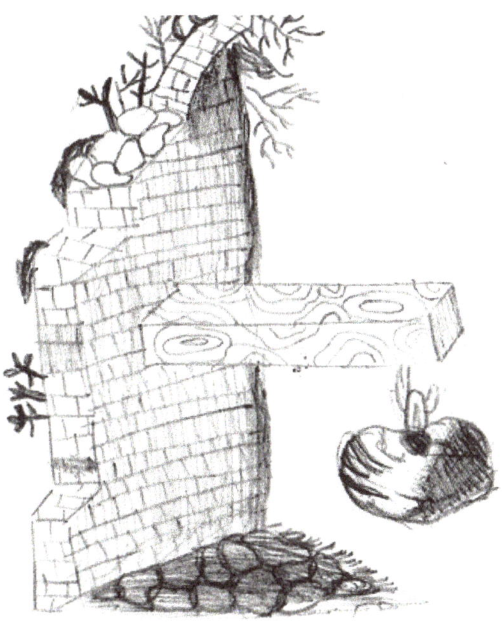

Figure 2.3.1 Galileo's Cantilever Representation

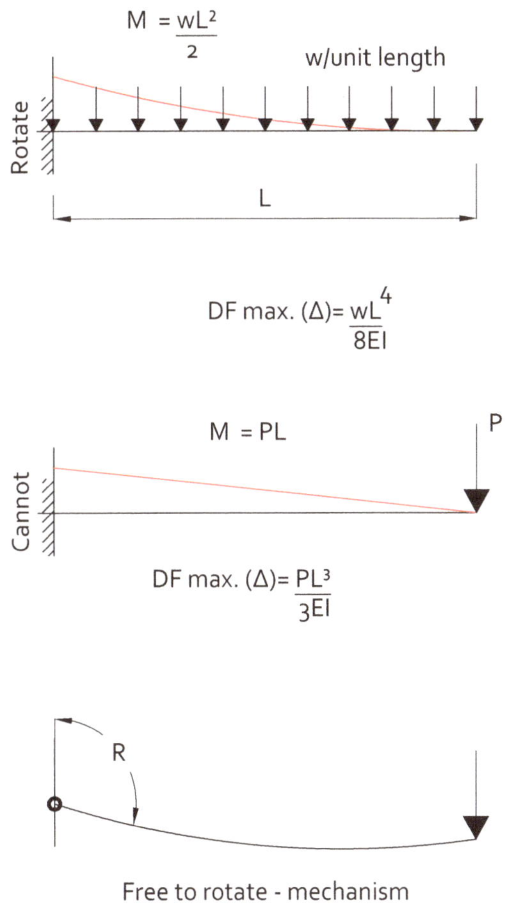

Figure 2.3.2 Cantilever Beam Principles

What distinguishes a fixed-end cantilever from a simply supported beam? It has a free end. The supported end is held up, but unlike the simple support, the beam cannot rotate. **Figure 2.3.2** shows that it would not work as a cantilever if the supported end could rotate.

Simply Supported Versus Cantilevered

Simply supported beams are structurally more efficient than cantilevered beams. Differences in the behaviour of a simply supported beam are illustrated in **figure 2.3.3**. Note that the bending is *hogging* as opposed to *sagging*. Tension strains occur on the top of the beam.

The deflection of the cantilever beam at the tip is **10** times that of the simply supported beam and the bending moment is **4** times greater in the same span cantilever.

Why Adopt Cantilevers?

Cantilevers are popular with architects. They can create dramatic façades, which are commonly derived from the need to provide upper-storey accommodation over limited space on the ground floor. Sir John Deanes College, shown in **figure 2.6.24** is a good example of a bold architectural statement for the school entrance using cantilever construction.

Bridges are often constructed by cantilevering out from bank-side abutments to form support for an internal span. This can avoid or minimise the need for building support piers underwater or in a deep ravine to support internal spans. A famous example is the Forth Bridge in Edinburgh (**figure 2.6.12**), which is built with cantilever towers in lattice steelwork construction supporting internal simply supported spans.

Balconies to buildings are often cantilevered to avoid supporting columns, which might be undesirable at ground level.

Cantilevering entrance canopies to buildings can provide shelter while at the same time signalling the entry point to the building; a statement for a corporation, perhaps.

Signage over motorways are cantilevering structures, as are many other examples.

To an engineer, cantilevers provide a challenge with increased shear, bending and deflections at the support compared with simply supported structures. Means of support, whether by solid supports in thick walls, back-spans or posts, all need to be validated; often required to be as slender as possible or hidden within a building's external walls.

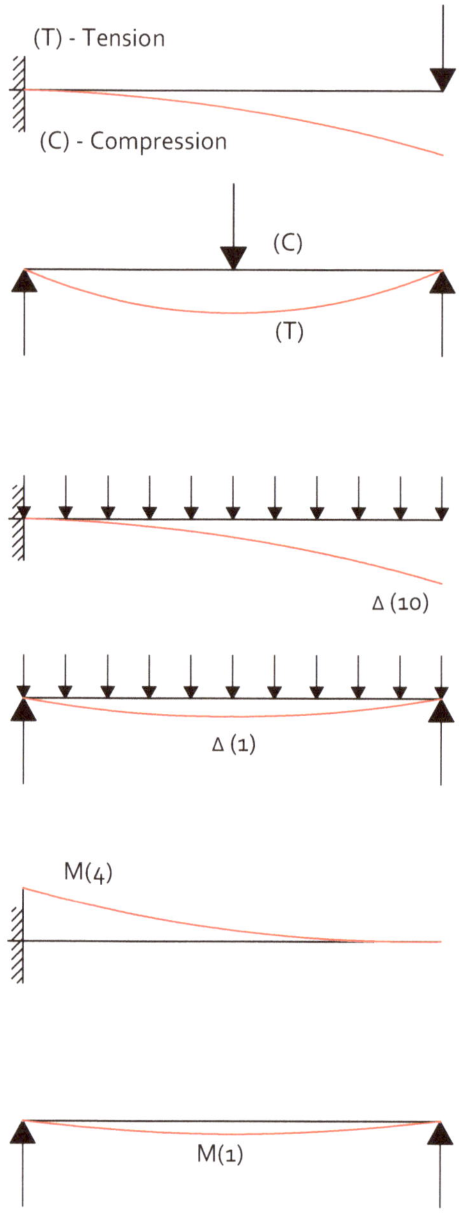

Figure 2.3.3 Cantilever Versus Simply Supported

Cantilever With Back-span

A common error which newly employed graduates fall into is to underestimate the deflections of beams, which comprise a simply supported condition with an overhang (generally termed cantilever with *back-span*). A cantilever with a *back-span*, in **figure 2.3.4**, will deflect more than an equivalent span cantilever which is solidly built into a support.

Figure 2.3.5 illustrates how an increase of back-span increases the deflection for a given beam size. Dimensionless deflections are shown to the left. The top figure shows the deflection with a purely built-in cantilever. The following diagrams show how increasing the back-span influences the tip deflection (and also the corresponding uplift in the centre of the back-span).

Consider the case of an engineer assuming that a beam is a pure cantilever when in reality, it has a back-span 4 times longer than the cantilever. The engineer will have underestimated the deflection by a factor of 5. In **figure 2.3.5** the relative deflections are illustrated non-dimensionally. With a back-span equal to the cantilever, the deflection doubles compared with a pure cantilever (top of figure).

Aside

Natural Frequency

With respect to the 'bounciness', the greater tip deflection reduces the natural frequency of the tip system (increasing the chance of resonating movements), but the added dead weight will tend to reduce the 'response' or 'user perception of movement'. This phenomenon will be discussed in the chapter, referring to vibration and sway. See chapter 2.12. (Figure 2.12.4).

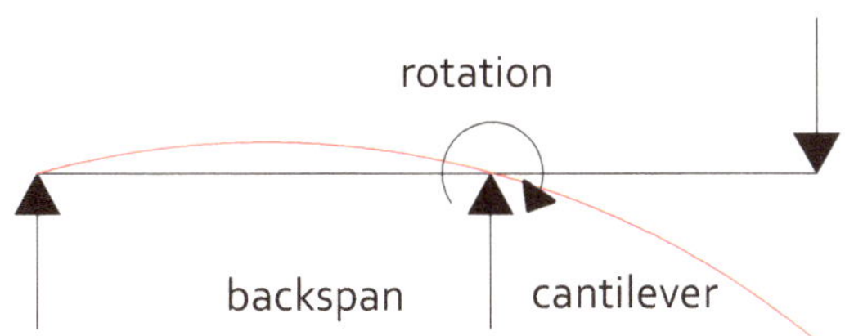

Figure 2.3.4 Cantilever with Back-span

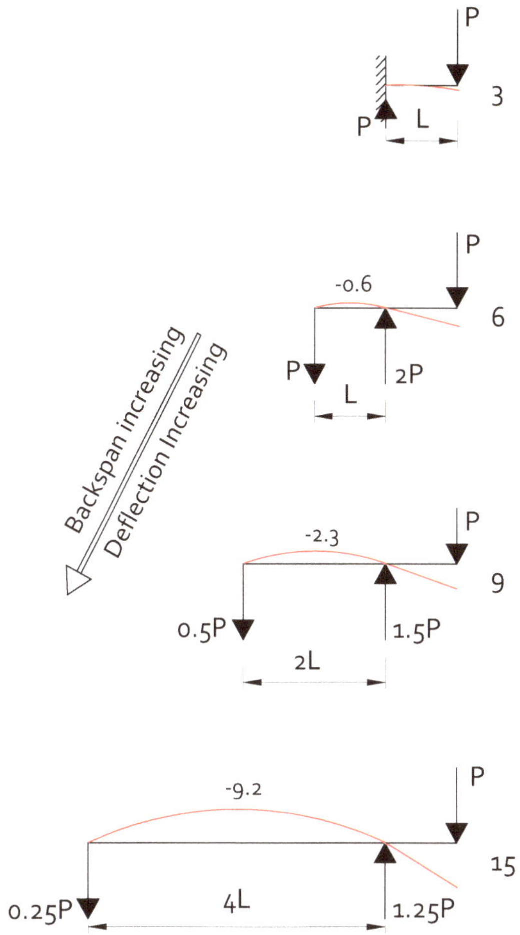

Figure 2.3.5 Cantilever with Increasing Back-span

Effect of Dead-weight on the Back-span

If a back-span is loaded, this has the reverse effect of cantilever loading. Sufficient ballast or holding down weight can be engineered to balance the tip loading. Referring to the previous illustration, a further complication to consider is the degree of loading on the back-span (the self-weight of the system and any live loading).

Figure 2.3.6 shows how loading on a back-span has the effect of trying to lift up the tip of a cantilever, the reverse of loading on a tip

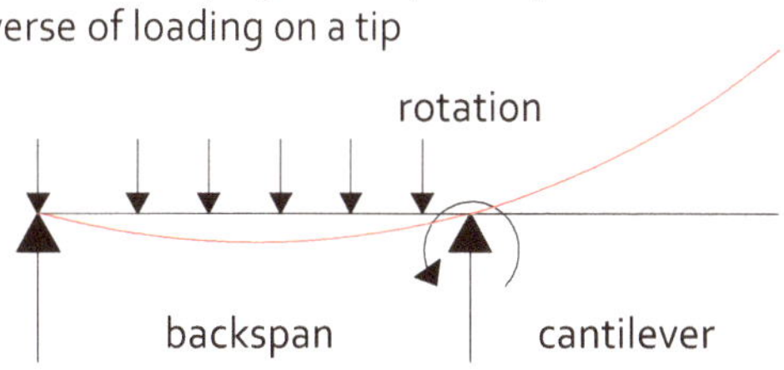

Figure 2.3.6 Effect of Dead-weight on Back-span

A glance at **figure 2.3.7** shows how a back-span that is greater in length than its cantilever will have a tendency to uplift the tip; the greater the deadweight, the greater the uplifting effect, which counteracts the tip deformation. This would be significant, say, in a situation with a heavy concrete floor with a lightweight tip load.

Varying Back-span with Dead-weight

Figure 2.3.7 iis based on the previous example, but a dead loading has been added along the length of the beam.

Deflections are shown non-dimensionally to the right. When the back-span increases, the uplift effect becomes significant at the tip.

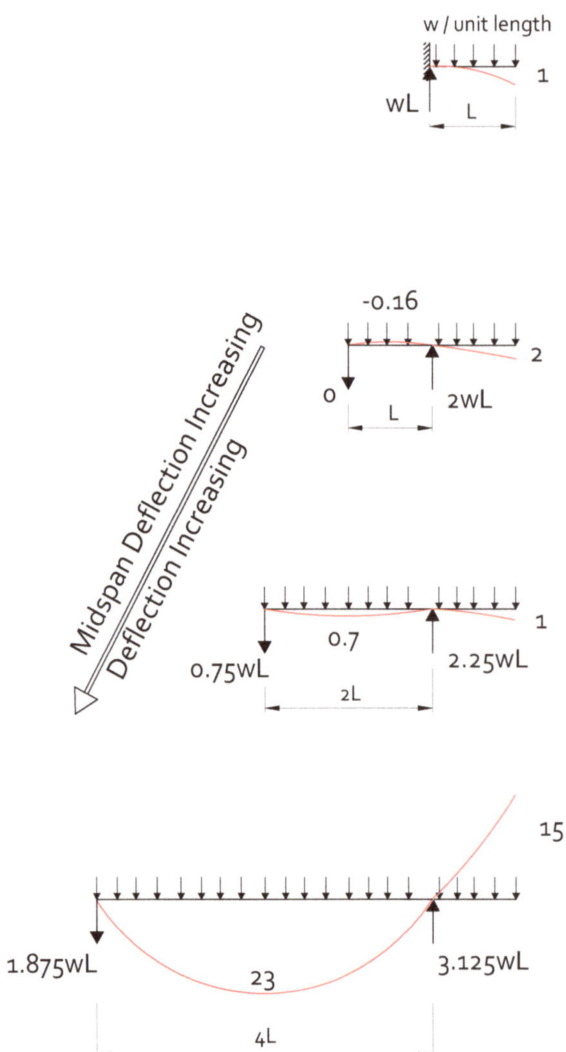

Figure 2.3.7 Variable Back-span with UDL

Approximate Loading Examples

Similar to the simply supported exercise in **figures 2.2.7** to **2.2.13**. **Figure 2.3.8** shows three similar load conditions. Note that the UDL (P/L) gives worst-case deflection (26).

Figure 2.3.9 shows a UDL substitute for a triangular load; substitute UDL gives a higher bending moment. However, the UDL support shear is lower (WL/3). Care should be taken if shear is critical.

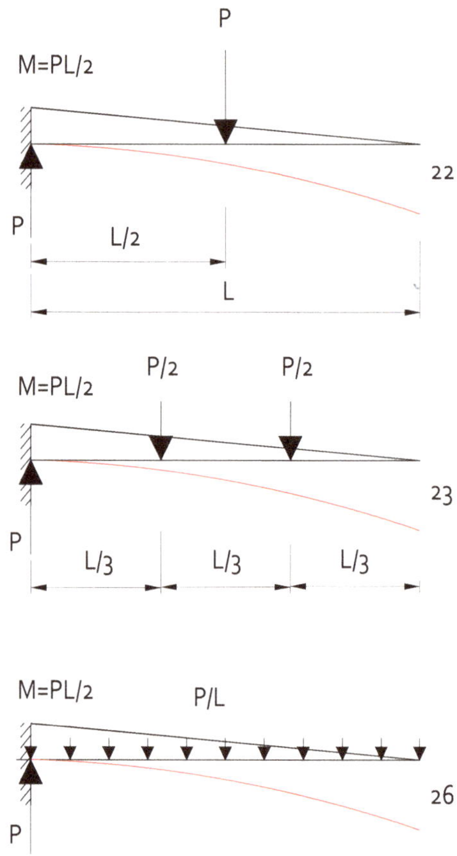

Figure 2.3.8 Interchangeable Point Loading

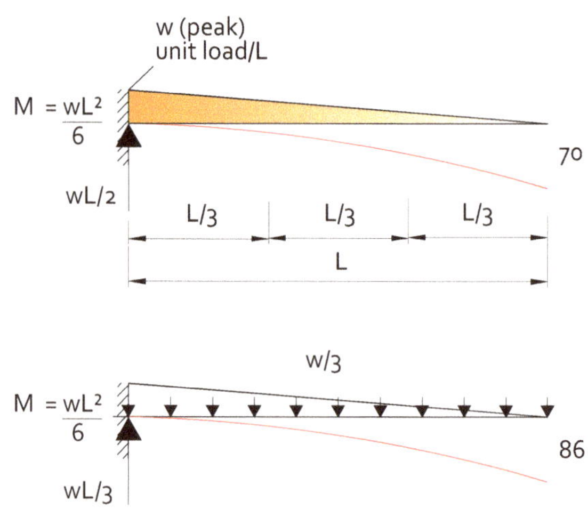

Figure 2.3.9 UDL Substitute - Triangular Load

Different Forms of Cantilever

With a few fundamentals, some real engineering solutions may start to be examined. Architecture, more often than not, limits the standard engineering solutions. Far from being a restraint on engineers, there is scope for ingenuity.

Two examples are shown alongside. The first is an office building, which has retail on the ground floor, as shown in **figure 2.3.10**. The shops are actually selling merchandise for Chelsea Football Club. In the retail business, high ceilings are at a premium; the higher the ceiling, the greater the visual expansiveness of the interior, and the greater the potential for the display of goods. In this building, the first-floor offices were maximised in the area by cantilevering over the ground-floor retail section.

A similar arrangement has been used on a smaller scale to support external balconies in existing houses. See **figure 2.3.11**. Balcony beams are fixed directly to posts, which may be built into the external wall cavity. The principle is the same as noted above for the office floors, with the posts resisting the bending moment from the balcony. There is a pulling-out effect at the eaves level; the column must be tied back into the house (the house must be capable of resisting this tie-back force from first floor to roof level).

Balcony Design

Balcony design is reviewed in **chapter 2.9** in which combined structural action is examined with regard to limiting the movement of a handrail. See **figure 2.9.6** to **2.9.9**

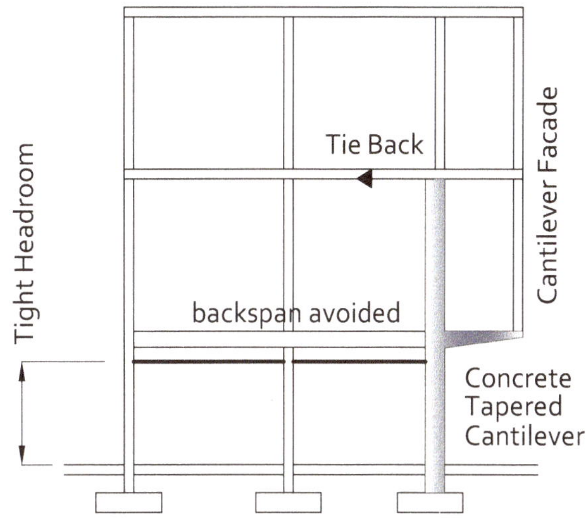

Figure 2.3.10 Overhanging Offices No Back-span

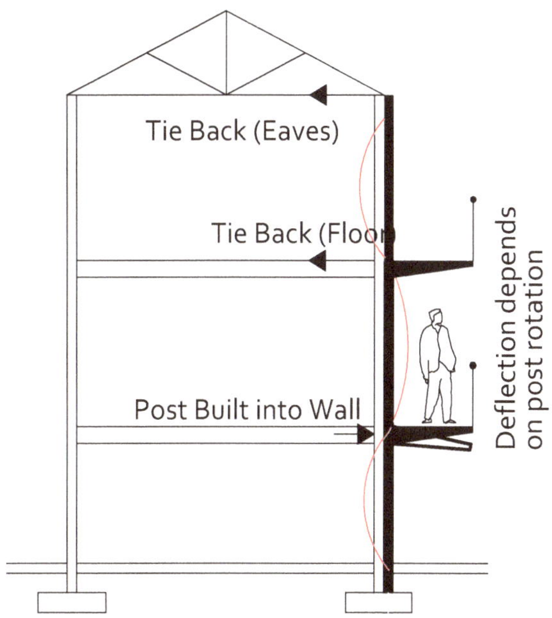

Figure 2.3.11 Cantilever Balcony

Cantilevering on a Larger Scale

There are very large-scale cantilevering structures – notably the cellular reinforced concrete beams which support the upper stories of the NatWest tower in the City of London illustrated in **figure 2.3.13 and 2.3.14.**

Figure 2.3.12 illustrates ways of cantilevering floors above ground level on multi-storey buildings. The whole weight of the cantilevered upper floors may be supported on a heavy cantilever with back-span, on each floor – with back-spans into the building, or using solid walls or plates.

Description of examples 1-4

(top to bottom) in figure 2.3.12:

1. **Cantilevered from first floor-solid slab or beam**
2. **Cantilevered from first floor-truss over floor depth**
3. **Cantilevered from all floors -e.g. NatWest Tower (figure 2.3.13)**
4. **Solid walls – concrete or plate**

Multi-storey Cantilevers

This form of structure for multi-storey buildings is commonly adopted in dense urban areas. The overhang is eye-catching. Furthermore, it allows wider pavements for pedestrian use around buildings, at the same time maximising upper floor usable area.

Whichever options in examples 1 to 4 in figure 2.3.12 are selected will depend on a number of factors. For example:

- Option 1 and 2 allow rapid construction with minimal floor depth over the thick transfer deck on first floor. Post-tension flat concrete slabs may be adopted for the upper stories.
- Option 3 standardises the upper floors. However, flat slab construction may not be appropriate owing to the need for thicker construction if the cantilevers become large.
- Option 4 is clearly only really appropriate for lift shafts and stair cores in local areas of the building.

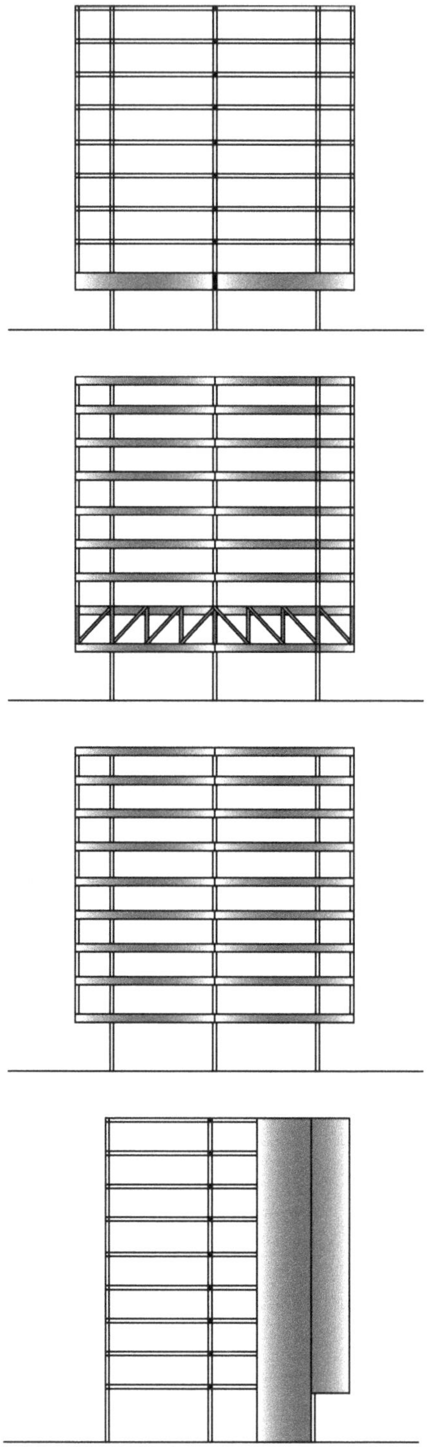

Figure 2.3.12 Multi-storey Cantilevers

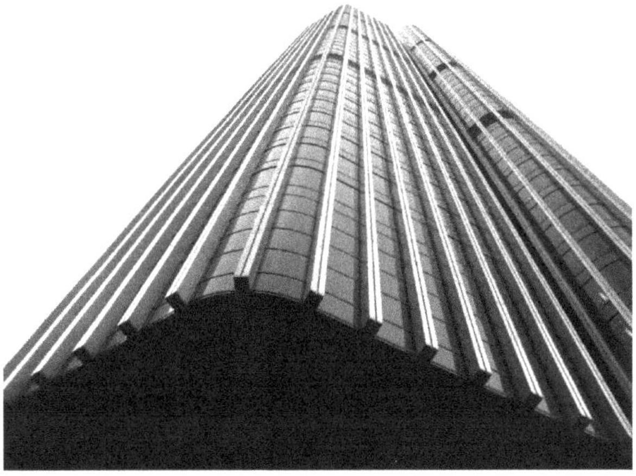

Figure 2.3.13 NatWest Tower (Tower 42)

NatWest Tower (Tower 42)

At 183m height, Tower 42, as it is now called, is one of the tallest buildings in the City of London commercial district. Each of the 47 floors cantilevers from a concrete core.

Under the first floor, rights of light are maintained at street level and better pedestrian access is possible.

Designed by Colonel Sieffert (architect) and Pell Frischmann (structural engineers), it has held its own since the 1960s on the London skyline.

In modern terms, the percentage of structure to usable floor area is not economical. It is unlikely that a similar solution would be chosen today except for special reasons.

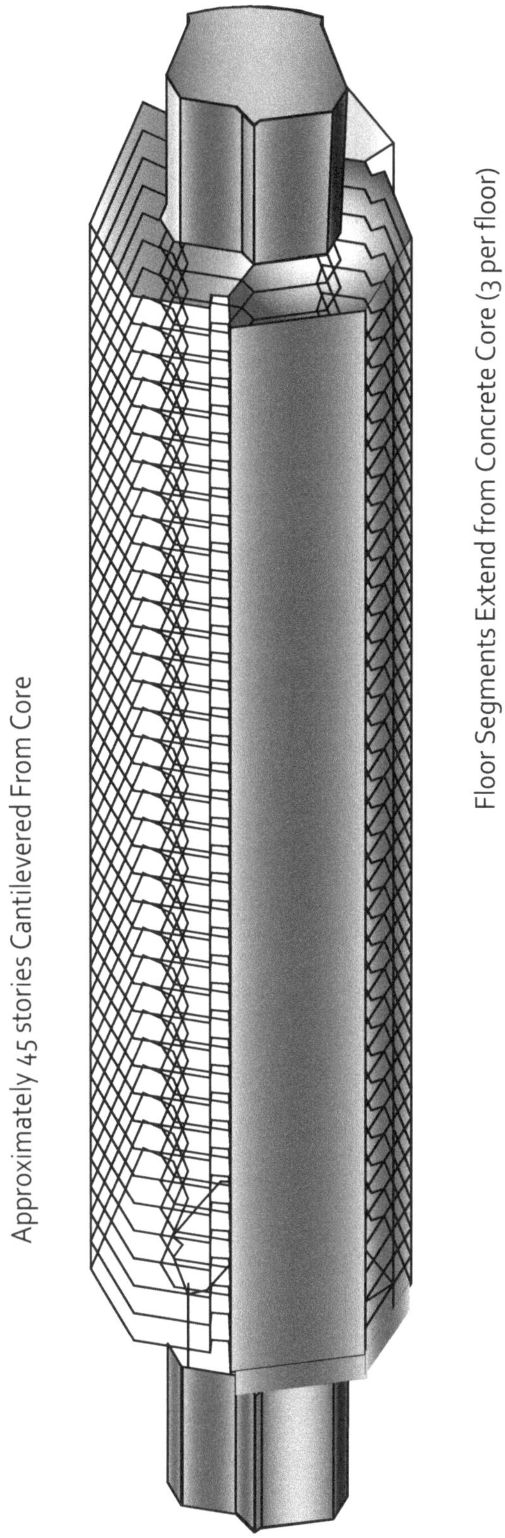

Figure 2.3.14 NatWest Tower Structure

Chapter 2.3
Quick Note Summary

1. Cantilever beams are more flexible and generate higher bending moments than the equivalent simply supported span (bending moment 4 times greater and deflection 10 times greater).
2. Cantilevers with back-spans have higher induced deflections than the equivalent purely fixed supports owing to rotations in the back-spanning beams.
3. A shorter back-span is stiffer, but the negative reaction (or uplift) is proportionally greater. The designer needs to check that there is enough weight in the structure behind to hold down the back-span.
4. Approximations for pattern point and triangular loading may be substituted with equivalent UDL loading.
5. It is possible to cantilever, for instance, overhang a floor without resorting to a back-span solution. This is achieved by fixing the beam to a stiff column or post built into the outer edge of a building.
6. With regards to 6 above, care needs to be taken to check the additional rotation caused by the post (i.e., deflection at the tip of the cantilever) and strapping back into the building of the post at the head. The post stiffness needs to be tuned to ensure the tip deflection is no more than approximately 6mm under dead load deflection for a balcony or floor. The strap detail at the head should be sufficient to tie back the reaction at the head of the post. The tie force also needs to be transferred via the floor diaphragm to the stability frame or walls.
7. On a grander scale, using the whole depth of an upper floor or multiple floors allows deeper structure to cantilever where greater loads are present, using a truss, deep beam solution, or wall, for instance

2.4 Continuous Beams

Add one internal support - or more- to a simply supported beam, and the resolution of forces becomes statically indeterminate. In the spans, the beam becomes stiffer than its simply supported counterpart. The beam bends over the supports (hogs) in an inverted form of the simply supported moment. The chapter follows with a brief examination of two and three equal-span continuous beams; a review of continuous unequal spans; the effects of asymmetric loading and differing adjacent beam stiffness; a useful method of adding a hinge in the spans to achieve statical determinacy; redistribution of moments from support to internal spans and the elastic shear effect in frames.

Many built structures are continuous, in particular steel and concrete framed structures. Design methods for concrete frames are mostly based on continuity - in beams and slabs - taking advantage of fixed connections by placing additional reinforcement over posts and columns.

Supports may be arranged under lengths of steel or timber to form continuity. When joints are needed, connections have to be designed to incorporate bending moments where beams meet at support locations. This can be avoided whilst attaining the benefits of continuity, as discussed later in the chapter.

In the examples that follow, **Δs** indicates *deflection*, **SSBM** indicates *bending moment*, and **SSR** indicates the *support reaction* of a simply supported beam with a span **L.** **'w'** is a linear load (*expressed commonly as kN/m*).

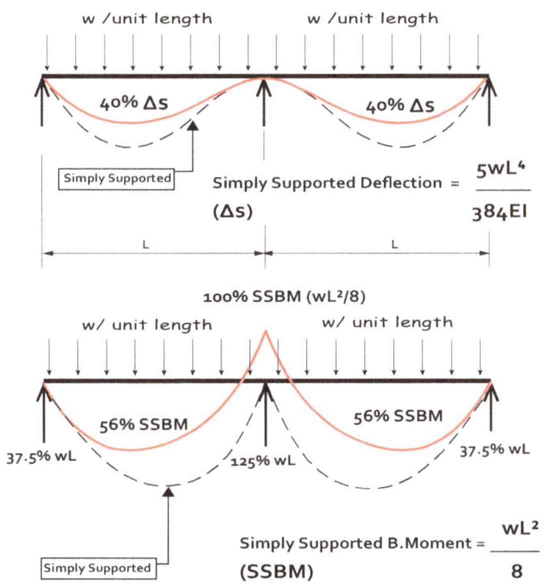

Table 2.4.1 2 Span Continuous Beam Equal Spans

2 Span Continuous Beams

Figure 2.4.1 below shows a simple continuous construction comprising a two-span beam with equally sized beams and equal loading on both spans. Internal forces and deflections are shown as a multiple of the equivalent simply supported span.

Table 2.4.1 compares the bending moments, reactions, and deflections, showing the increased stiffness of the 2 span continuous beam compared with a similar span simply supported beam.

Equivalent UDL Comparison	
Simply Supported 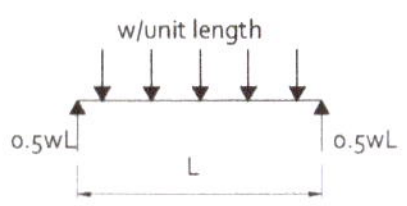	2 span Continuous 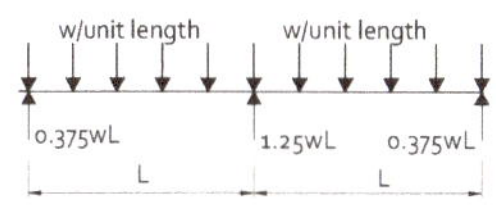
Bending Moment Midspan	$M = \frac{wL^2}{14}$ 56% Simply Supported
Shear Force Supports	$V_{end} = 0.375wL$ 25% less -end $V_{mid} = 1.25wL$ 25% greater -middle
Deflection Midspan	$D = \frac{2wL^4}{384EI}$ 40% simply supported

Table 2.4.1 Simply Supported 2 Span Continuous Compared

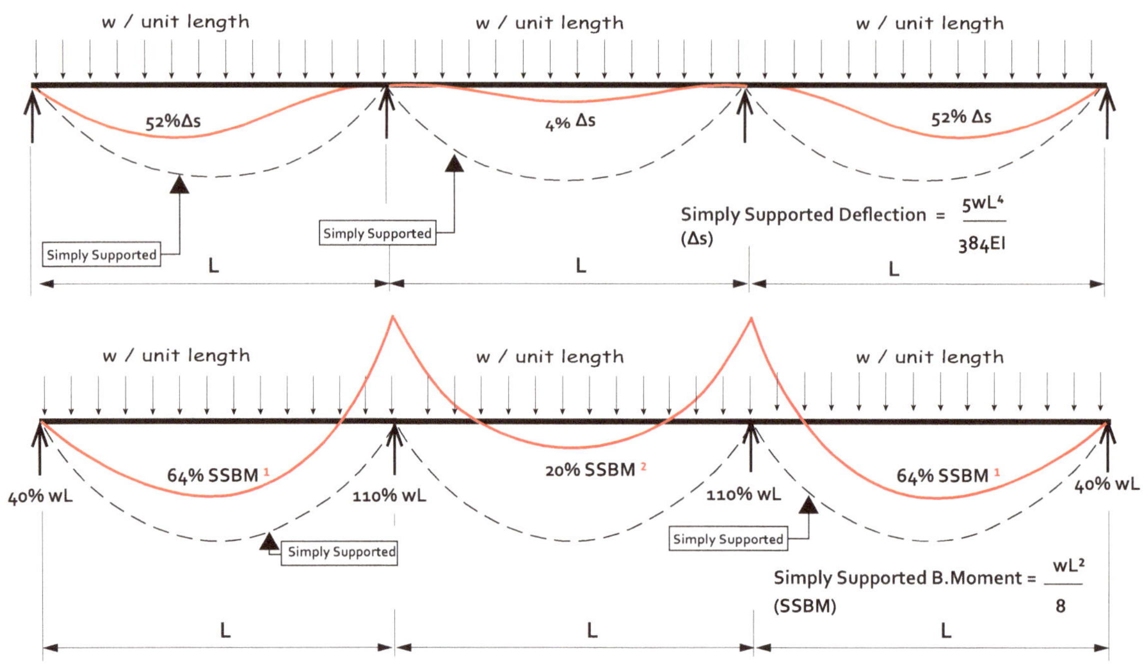

Figure 2.4.2 3 Span Continuous Beam Equal Spans

3 Span Continuous Beams

By adding another continuous span, this smooths out the peak values compared with a 2 span beam. **Figure 2.4.2** illustrates the bending moments and deflections as a proportion of the simply supported case, as with **figure 2.4.1** .

Resultant bending, shear, and deflection are shown as a proportion of the equivalent simply supported span in **figure 2.4.2**. Further economy of structure may be attained by the addition of a third span.

Footnotes: Figure 2.4.2

Bending moments may be otherwise expressed as:

[1] $wL^2/12.5$

[2] $wL^2/40$

2 Spans Unequal Length

Figure 2.4.3 is an illustration of the effect of unequal spans. The second span is half the length of the first.

An appreciation of the effect of altering the relationship of spans comes into practical use when deciding to provide intermediate supports to a long beam in a building; the span ratio of adjacent spans may determine the selection of support points.

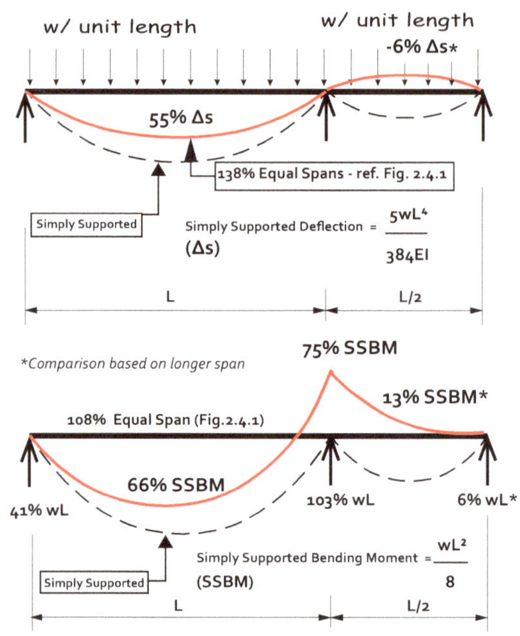

Figure 2.4.3 Influence of Adjacent Spans Both Spans Loaded

Comparing with **figure 2.4.1** with like-for-like loading the beam in **figure 2.4.3**, in summary, shows an increase in mid-span deflection of the longer span (38%). The most significant influence on the mid-span deflection is the fact that a shorter adjacent span has a *smaller counterbalancing* moment owing to the reduction in overall weight alongside a bending moment in the central support (which generates reverse rotation).

Pattern Loading

Returning to two equal spans, an important design consideration in structural design is that of unbalanced loading. **Figure 2.4.4** indicates the effect of loading only one adjacent span. Comparing **figure 2.4.1** with **2.4.4**, the absence of the right-hand loading increases the (LHS or left-hand side) mid-span moment.

With a little arithmetic, it can be seen that the moment in the equally loaded 2 span beam is wL2/14, which increased to wL2/11 when the loading is removed – *approximately* 30% increase.

The bottom part of **figure 2.4.4** shows graphically how the upward deflection of the RHS span (or mid-span deflection) may be calculated by means of a free-body diagram. Cutting off the LHS span requires a restoring moment. Equations are at hand, which relates the mid-span deflection of a beam with 2 pin supports to an end-applied moment (the value is rounded up).

Pattern loading is generally used in practice to define maximum moments where live loading is applied in different *on/off* patterns to assess the worst-case beam moments. Dead load is combined as a constant with the live pattern loading. The outline of worst-case bending moments is often referred to as an *enveloping case or blanket case.*

Pattern Loading with Unequal Spans

Taking the concept of the previous section further, an investigation into the effect of unequal spans is summarised in **figure 2.4.5.**

The percentage differences in the bending, deflection, and shears compared with **figure 2.4.4** are indicated alongside.

Comparing with **figure 2.4.4** with **figure 2.4.5**, if a short span is unloaded, the mid-span deflection and bending on the RHS is some 88% lower than for equal spans.

Since the shorter span is relatively twice as stiff as the longer span, it raises the support moment (compared with equal spans). In turn, this *pulls up* the mid-span moment on the LHS.

Loading Approximations

In **Chapter 2.2**, with respect to simply supported beams, a number of approximated loading conditions were considered. See **figures 2.2.7** to **2.2.13**. Use of these approximations may be adapted to continuous beam checks if the actual loading is irregular or triangular.
In this case, the simplified loading pattern may be substituted using most of the examples in this chapter, which are based on uniform loading or central point loading.

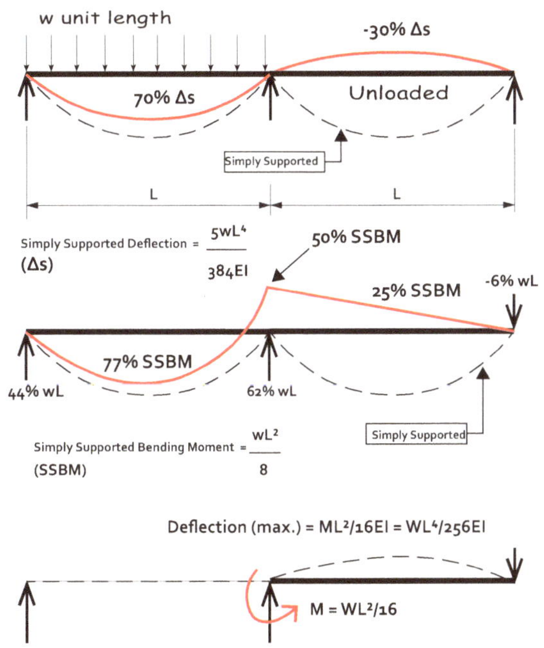

Figure 2.4.4 Influence of Adjacent Span One Span Loaded

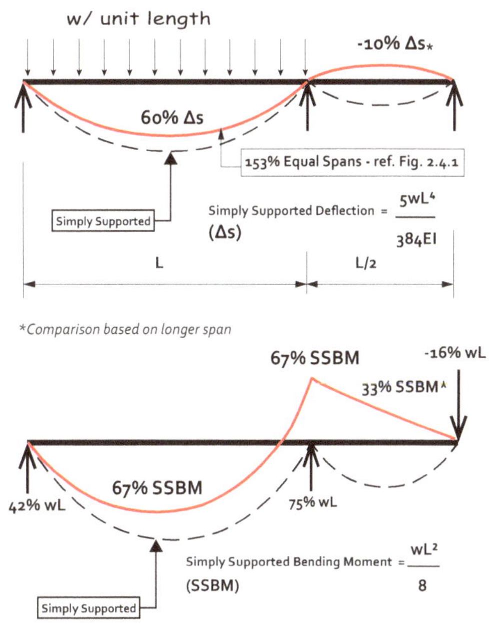

Figure 2.4.5 Influence of Adjacent Span Varying Span

Aside

Moment Distribution – Hardy Cross Method

In 1930 Professor Hardy Cross of the University of Illinois wrote a paper: "Analysis of Continuous Frames by Distributing Fixed-End Moments".

This method became a universal tool for hand calculation of continuous beams and frames. Essentially, it is an iterative method whereby assumed fixed end moments – set up for individual beams according to their loading - are then redistributed according to the relative stiffness of adjacent structural elements that meet at a fixed connection. The process is repeated until it converges; the solution then allows the calculation of shear forces and deflection estimates.

'Understanding Structural Analysis' (Brohn 2005, ref. 1. Chapter 9.) covers this method succinctly. It is a method suited well to spreadsheets and a good means of attaining an understanding of the behaviour of statically indeterminate structures, although it is less often used owing to the widespread adoption of structural analysis software.

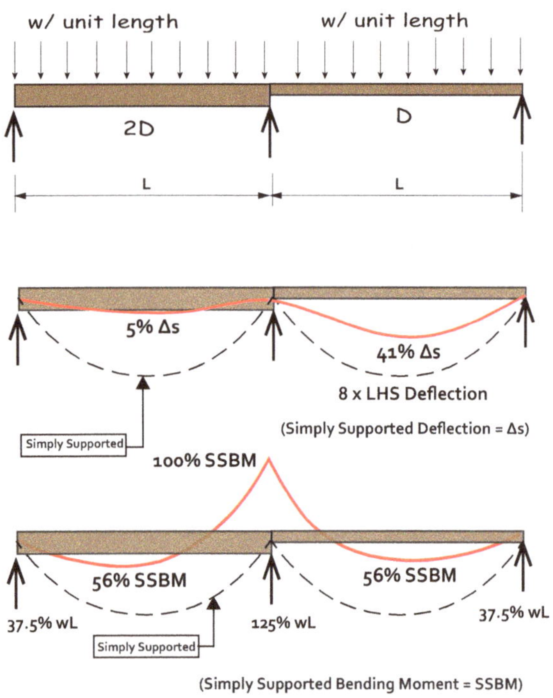

Figure 2.4.6 Influence of Adjacent Span Varying Stiffness

Adjacent Spans with Varying Stiffnesses

On occasion, continuous beams vary in stiffness along their length. The deflections, bending moments, and shears are influenced by this. A simple example is illustrated in **figure 2.4.6**. Two equal spans, which are equally loaded, form a continuous beam (fixed connection at the central support). The left-hand span (LHS) is a rectangular section with depth D, and the RHS similar with depth 2D.

Calculating the moment of inertia (I) for each reveals that the I (inertia) value is proportional to the cube of the depth – i.e. the LHS is 8 times stiffer than the RHS.

Transfer of moment across supports is proportional to the I/L ratio (Inertia/Span). The bending moment diagram is symmetrical, which is a slightly surprising result, given the inequity of stiffness.

Stiffness on 2 equal spans with equal loading on closer inspection reveals that the central bending moment is the same as for 2 equal stiffnesses. Distribution of the external moments, by calculating as per **figure 2.4.5.**, is independent of the relative stiffnesses.

The support reactions are externally applied forces related to the distribution of moment. In other words, the support reactions are the same as for the symmetrical case illustrated in **figure 2.4.1**

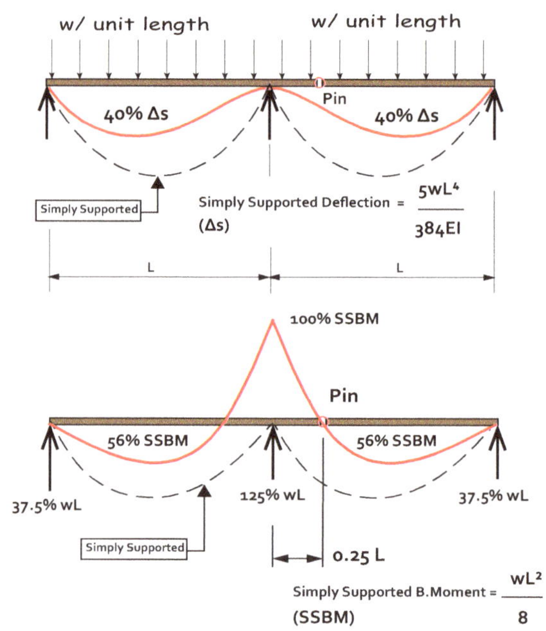

Figure 2.4.7 2 Span Continuous Beam Hinge at 25% L

Strategic pins – Continuous Beams with Statical Determinacy

Pin at Quarter Span Position

An indispensable tool for engineers is to utilise the properties of a continuous beam to increase the effective stiffness of a span without having to supply or build a beam in one single length.

Figure 2.4.7 is a 2 span beam which - to all intents and purposes - is almost identical to the 2 span continuous beam with equal loading on both spans in **figure 2.4.1.**

A pin has been inserted at a quarter-span position to the right of the middle support. This location more or less coincides with the point of contraflexure in the two-span beam in **figure 2.4.1**; in other words, the point of zero bending moments. In such a manner, two beams may be effectively made into one continuous beam.

Pin at Third Span Position

Figure 2.4.8 ddemonstrates the effect of moving the pin away from the middle support to a position that approximates to the third span position. Note that, compared with the quarter span example in **figure 2.4.7**, there is an increase in the mid-support moment, a decrease in mid-span moments, and an increase in the 2nd span deflection.

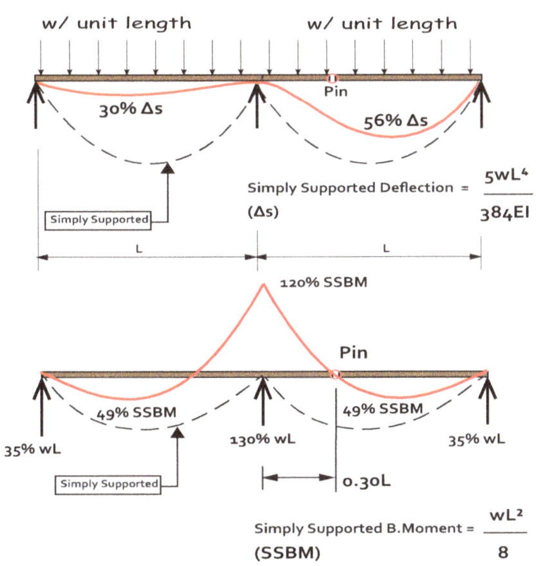

Figure 2.4.8 2 Span Continuous Beam Hinge at 30% L

Hinged Cantilever Example

A good place to see hinged cantilevers is on highways. Bridges that cross over the main highway are often built with the use of a hinged cantilever. The heavy supporting abutments may be built away from traffic (or on the banks of a river) to allow the central deck to be simply supported. This may possibly be fabricated offsite and lifted into position overnight when the road is closed. See **figure 2.4.9.**

In summary, the benefits of the hinged cantilever include:

1. They may be analysed as statically determinate problems.
2. Reduced deflections and mid-span moments compared with simply supported beams.
3. If two sections are used for a 2 span beam (this applies also to multiple spans), the connection between the two beams may be simplified to a hinge connection. This allows ease of design and fabrication over forming fixed connections – a particular difficulty for steel and timber structures.
4. Shorter beam sections may be used for ease of transportation.
5. Because of 2 above, shallower beam sections may be used over the equivalent simply supported condition. This may be particularly useful in buildings where headroom under beams may be critical and where multiple internal supporting columns may allow the use of continuous spans.

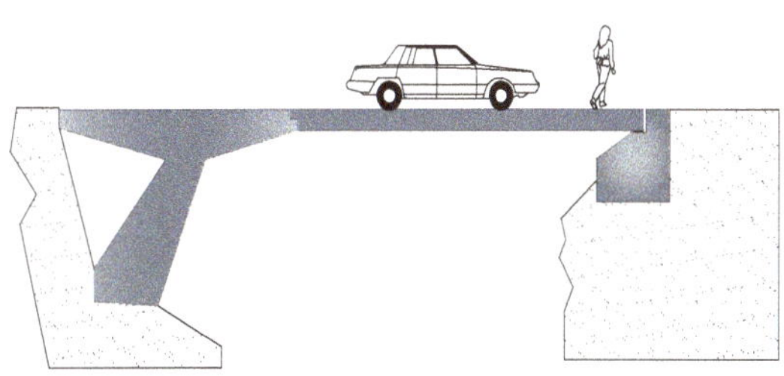

Figure 2.4.9 Example of Hinged Cantilever

Redistribution – Smoothing Out Peaks and Troughs

A process of redistribution is often carried out in concrete-framed building for continuous beams. This is a way of creating a more even arrangement of reinforcement. Referring to the continuous beam in **figure 2.4.1**.the support moment is wl2/8 and the mid-span moments are wl2/14. If the beam is allowed to go plastic or yield at the central support,, the bending moment in the mid-span may be reduced to approximately 70% of the standard case ref. **figure 2.4.1**., based on a guideline used in the former British Standard in the UK.

At the same time, to maintain equilibrium, the mid-span moments on either side will increase to the value shown in **figure 2.4.10** – an increase of 14% (see **figure 2.4.1**).

The effect of flattening out the peaks and troughs is to make the redistributed beam behave in a *transition state* between fully continuous and simply supported. Of note, the support reactions, shears, and deflections will alter.

A relatively simple method of analysis may be used by taking the middle support moment, which has been redistributed, and subtracting/adding the calculated shear and deflections in combination with the simply supported beam standard values.

A free body diagram, ref **figure 2.4.11**, shows how the reduced mid-support moment may be used to calculate the aggregate deflection and shears. The deflection and shear thus calculated may be subtracted from the equivalent mid-span moment and support reaction for a simply supported beam with UDL.

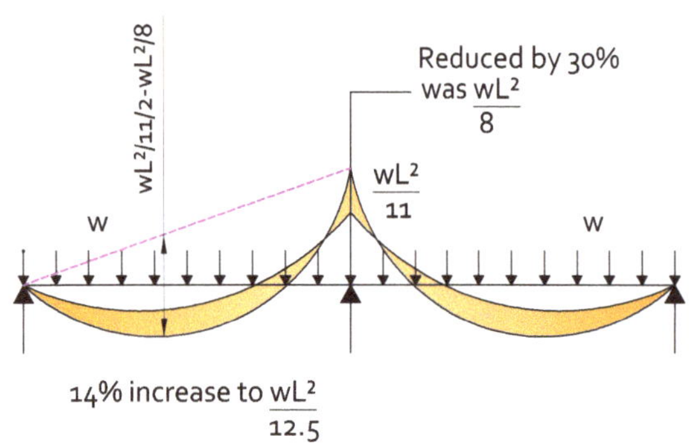

Figure 2.4.10 2 Span Redistribution Example

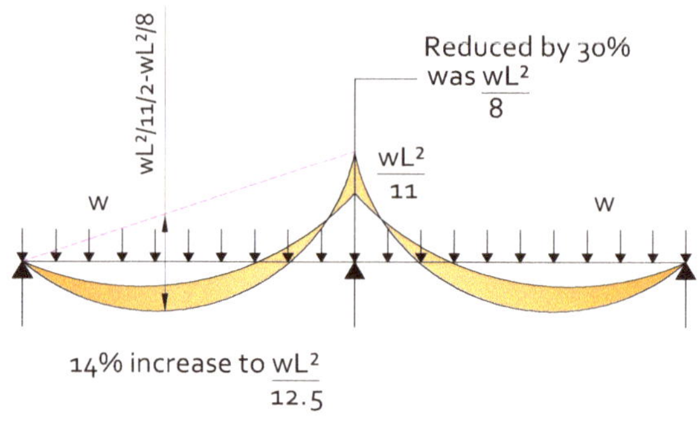

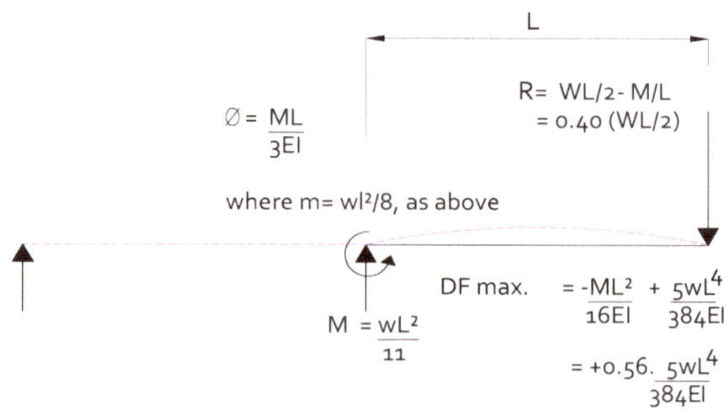

Figure 2.4.11 2 Span Redistribution Deflection and Reaction Calculations

Concrete Framed and Fixed Framed Buildings – Elastic Shears

Figure 2.4.12 highlights an important consideration that is sometimes overlooked.

As noted in this chapter, a function of continuity is that it increases the internal reactions on a two-span beam up to 25% over that which would be expected for a simply supported beam.

Where column and beam connections are designed to be rigid in a building rather than pinned, the internal transfer of bending moments into the joints between columns and beams causes an *elastic shear effect*. **Figure 2.4.12** sshows reactions as a percentage of the full span load for a framework that comprises 6 equal spans.

The outer column has a reaction of 39, which is 22% less than a reaction that would have been calculated for a simply supported condition. The first internal support has an increased reaction 13% greater than the equivalent simply supported condition.

This effect is of particular note when calculating the loading on columns, generally named in the UK a *column load takedown*. If simply supported conditions are assumed for assigning the loaded areas to the column, then the first internal support loading will be underestimated.

Calculations for concrete framed construction generally take this into account even if the frame elements are designed as simply supported or with pinned connections owing to the inherent rigidity of concrete frames, and fixed connection computer models will adjust for this effect.

An engineer should always be aware of this phenomenon when considering column foundation loading to ensure that this effect is included in the calculations.

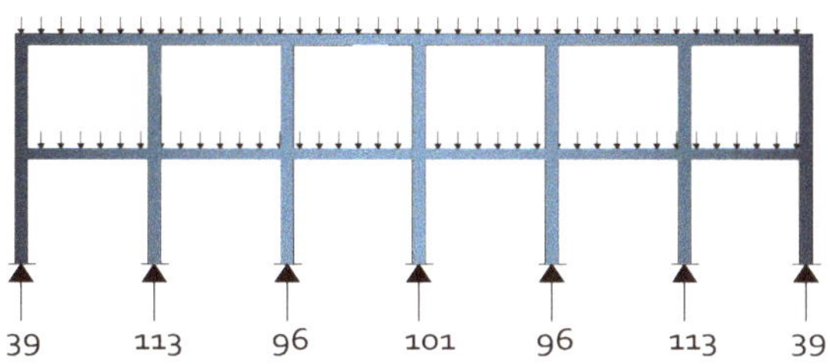

Figure 2.4.12 Fixed Frames Elastic Shear Effect

Avoiding the Elastic Shear Effect

In **chapter 2.2** the stored column problem is essentially a continuity problem (figure 2.2.4), which can create difficulties when supports are not infinitely stiff.

Figure 2.4.13 illustrates a series of ground beams that are on piled foundations. The piles were designed with some settlement in soft ground expected. For an efficient design the ground beams would have been designed as continuous over the piled foundations.

Predictability of pile loading was an important

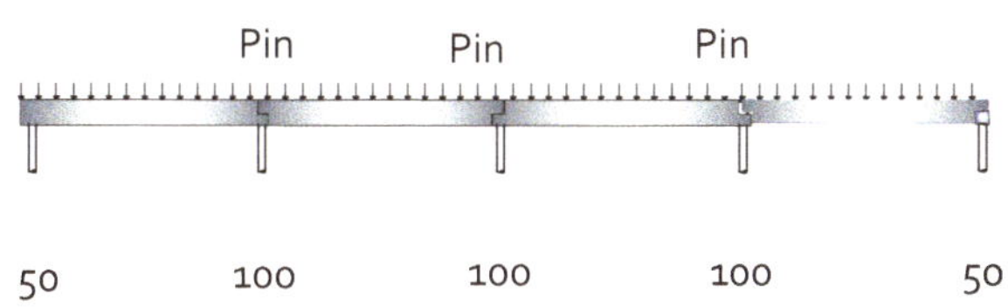

Figure 2.4.13 Avoiding the Elastic Shear Effect

consideration in this situation. To enable predictable pile loading, the beams were decoupled with pinned connections over the pile supports. Since there were no continuity moments over the supports, the structural system was viewed as a series of adjacent simply supported beams.

Although the beams were not as efficient as they might be in the section, the piles could be designed with confidence in their predicted loading. It is sometimes useful to revert to simple construction.

Point Loading

Point loads are similarly affected by continuity. **Figure 2.4.14** is an example of a centrally loaded span.

When compared with the simply supported beam internal forces and deflections (ref. **figure 2.2.3**), with an adjacent equal span, the deflection is 72% of the equivalent simply supported beam, and the bending moment at mid-span is 80% (PL/5) of the simply supported case.

Continuity will reduce the equivalent mid-span moments at the expense of the structure having to resist hogging moments at the internal supports.

Other examples with pattern loading and unequal spans may be similarly verified.

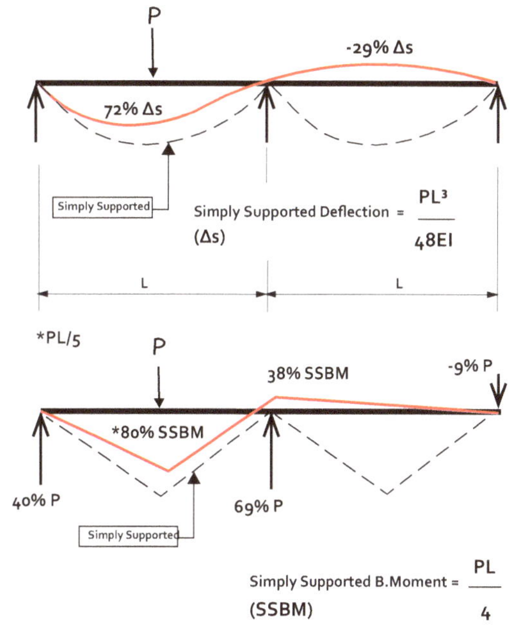

Figure 2.4.14 2 Span Continuous Beam Point Load LHS

Chapter 2.4
Quick Note Summary

1. Equally loaded 2 span continuous beams offer enhanced stiffness for adjacent spans and reduced mid-span moments. The mid-span support has an increased reaction (25%), and the mid-support moment is equivalent to the simply supported mid-span moment.

2. 3 span beams and other multiple spans have similar properties to varying degree. Internal spans exhibitthe further economy of bending moments and deflection.

3. Unequal spans have peculiar properties. A shorter adjacent span with loading the same as the longer span can have an influence such as increasing the bending and deflection of the longer span (compared with equal spans ref. **figure 2.4.1**).

4. Pattern loading is an important design consideration pertinent to continuous beam design.

5. Based on Chapter 2.2, figures 2.2.7-13 irregular or triangular loadings may be substituted using the appropriate factors.

6. Transfer of moments across internal supports proportional to the I/L ratio. If the inertia between adjacent beams vary, and even with the same span, the distribution of internal forces is affected.

7. By placing pinned connections strategically within adjacent spans but maintaining continuity over internal supports it is possible to achieve the effect of continuity and simplify construction at the same time.

8. The peaks and troughs of support and internal mid-span moments may be smoothed out by redistribution. By allowing plasticity at the support, mid-span moments may be enhanced to allow for a closer balance of internal forces, creating a more efficient section.

9. For fixed-frame buildings and continuous spans, the elastic shear effect should be carefully noted – the first internal spans are subject to increased shears at the supports.

10. Elastic shears may be avoided by releasing adjacent span bending moment transfer (i.e. use pinned supports). This can be used to ensure the predictability of support loading.

11. Point loading is influenced by continuity in a similar way to UDL. If a 2 equal span beam is loaded on the LHS (left-hand side) with a central point load, the moment and deflection in the mid-span of the LHS are 80% and 72%, respectively, of the simply supported case (ref. **figure 2.2.3**).

Chapter 2.5

2.5 Framed Structures

A fixed framework comprises structural beams and posts connected together with fixed connections transferring moments across the joists. Distribution of forces and deflected shape will depend largely on the degree of fixity of the supports (the extent to which moments are transferred). A recap of fixed-end moments is followed by a review of moment distribution in a three-member joint with varying support fixity. The principle of simple braced frames, wind frames, and goal post frames is explored, showing how simplifications can lead to fast and effective evaluation of frameworks.

Fixed End Moments

As a reference point, a fixed end beam will be examined, followed by a propped cantilever. Eliminating rotation, as summarised in figure 2.5.1, the midspan moments, support moments, and deflections are all significantly reduced when compared with a simply supported beam (ref. **figures 2.2.2** and **2.2.3**).

Another example that will help explain the frameworks described in this chapter is the propped cantilever.

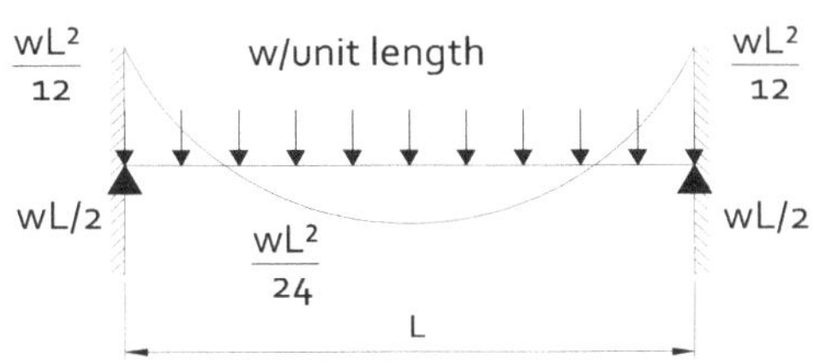

DF max. $= \frac{wL^4}{384EI}$ (20% Simply Supported)

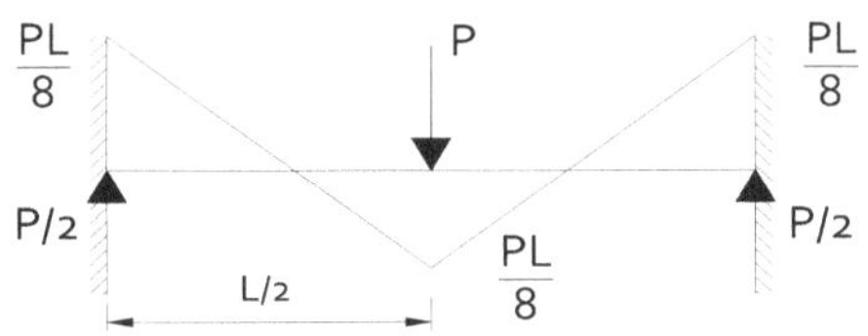

DF max. $= \frac{PL^3}{192EI}$ (25% simply Supported)

Figure 2.5.1 Fixed End Beam

Propped Cantilever

A propped cantilever may be viewed as the fixed end beam with a moment release (or permitted rotation at the RHS support). See **figure 2.5.2.**

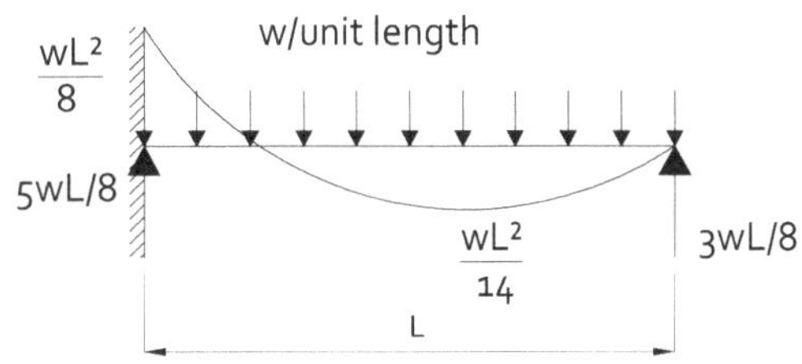

DF max. (0.58L) $= \frac{wL^4}{185EI}$ (42% Simply Supported)

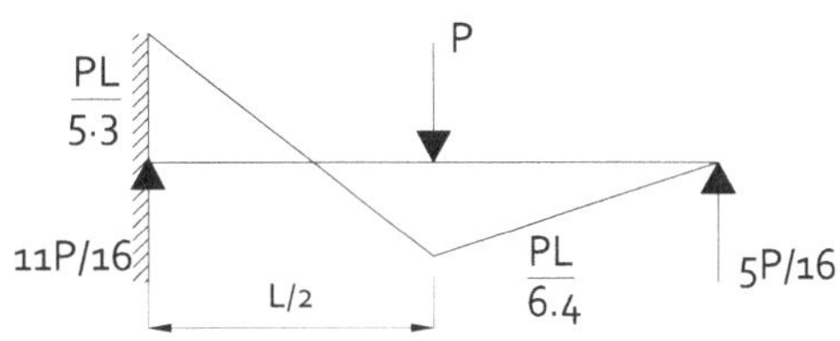

DF max. (0.55L) $= \frac{PL^3}{107EI}$ (44% Simply Supported)

Figure 2.5.2 Propped Cantilever

The two examples of the fixed end beam and the propped cantilever in **figures 2.5.1** and **2.5.2** are a clear illustration of how releasing the rotation restraint at one end of a beam has a significant effect on the internal forces and the deformation of the structure.

3 Member Frame Fixed Supports

A three-member frame with identical elements (length and section) with fixed-end supports was analysed. See **figure 2.5.3.**

A dimensionless moment of 54 was applied at the central joint between the beam and column.

As a result, the relative stiffness in flexure of the beams being equal, the bending moment is split into three equal portions into the posts and beam.

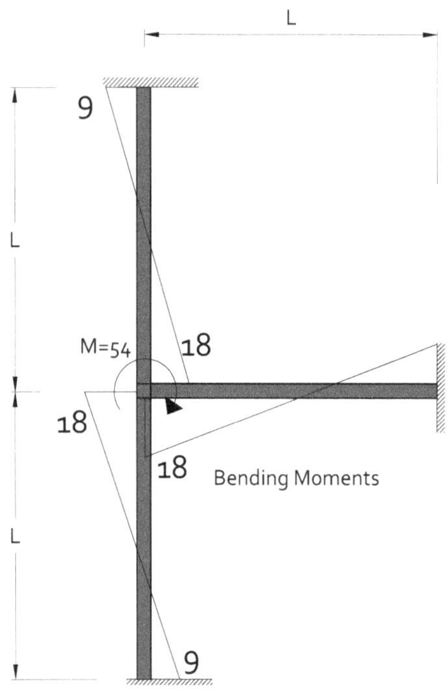

Figure 2.5.3 Three Member Frame Fixed Ends

Aside

Hardy Cross Method

Returning to the method of moment distribution introduced in the previous chapter figure 2.5.3 shows how the principle of 'carry over' works. At the central joint, each beam has a bending moment of 18. Half of this moment is exerted at the support (9). In other words, 50% of the moment at the fixed joint is carried over to the support.

Moment distribution is a process of assuming fixed ends at each joint to produce moments that are incrementally released according to the conditions of support at each end of an individual beam. For those interested, this is explained in 'Understanding Structural Analysis' (Brohn 2005, reference 1. Chapter 9.).

3 Member Frame UDL on Beam

Fixed Supports

In **figure 2.5.4** a dimensionless UDL of 10 was applied to the beam. Converting the bending moment at the RHS support (97) to an equivalent ratio of circa wL2/10, this is a halfway house between a fixed end beam (**figure 2.5.1** - wL2/12) and propped cantilever (**figure 2.5.2** - wL2/8).

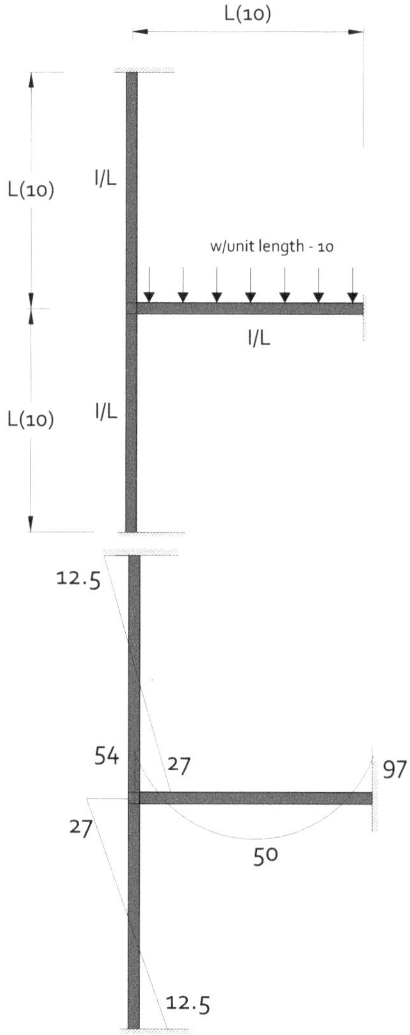

Figure 2.5.4 Three Member Frame Beam With UDL

Why is this? The columns to the LHS represent a *partially fixed connection*, supporting the beam as a prop but allowing partial rotation at the support. Any moment applied will be resisted by the columns, but having a finite stiffness will bend and allow twist in the section.

Further examination of the moment distribution demonstrates the midspan moment of 50 equates to circa wL2/20, which is a little greater than the fixed end beam midspan moment.

3 Member Frame UDL on Beam

Beam End Pinned (Released)

In **figure 2.5.5** a further release is applied to the RHS support to the beam – with the UDL of 10 applied to the beam. A way of looking at this problem is perhaps a *'propped not-quite cantilever.'*

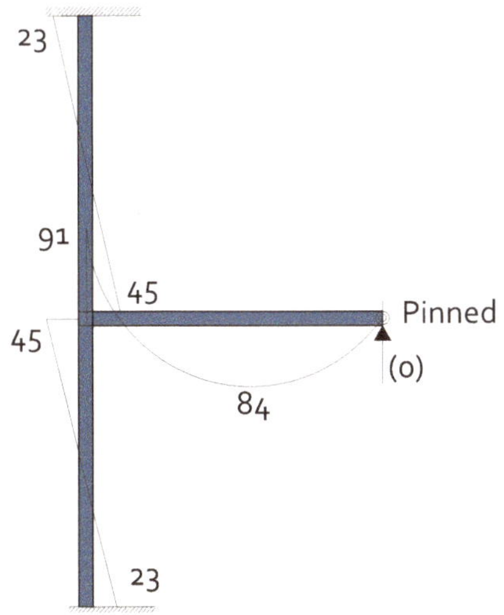

Figure 2.5.5 Three Member Frame Beam With UDL – Pinned End

Compared with the UDL in **figure 2.5.2**.LHS support moment is approximately 30% less than the propped cantilever case (**figure 2.5.2**).

Midspan moment is about 20% higher than the propped cantilever case. Midspan deflection is about 40% greater than the propped cantilever case.

See also below how the moment distribution alters compared with the previous fixed end beam case.

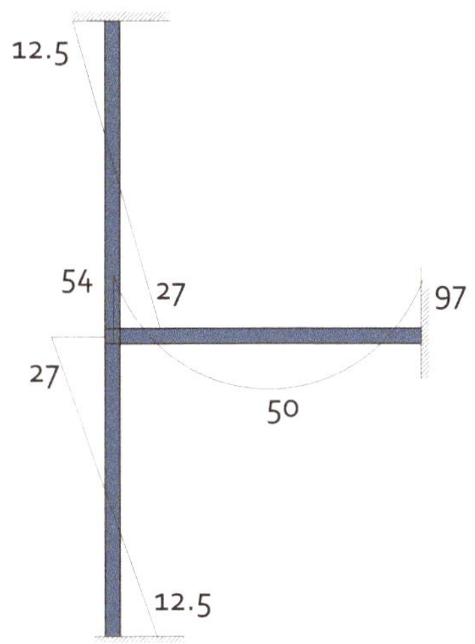

(Figure 2.5.4 – for side by side comparison)

Frames Fixed and Braced

Building frames are generally either fixed or braced. Fixed frames are a combination of posts and beams which are connected rigidly together. Transference of bending moments across the joints stiffens up the frame; the overall structural action depends mainly on the relative stiffnesses of the elements and the support conditions.

Figures 2.5.1 to **2.5.5** demonstrate how the support conditions are hugely influential on the distribution of internal forces within the elements in a fixed framework. Deflections are also greatly influenced by the support conditions and element stiffness. By altering the support conditions progressively, this is a good example of a trial-and-error approach to analysis.

Whilst the engineer might not have a full understanding of the internal forces by altering the support conditions and examining the internal effects judgements may be made about the structural behaviour of the real structure.

Purely fixed supports are rarely achieved in practice. Comparing the results of a frame with support as fixed and pinned, the engineer is able to make an assessment of what the *real* behaviour will be somewhere between the two conditions.

Braced frames are now compared with fixed frames. See **Figure 2.5.6** for an example of a braced frame.

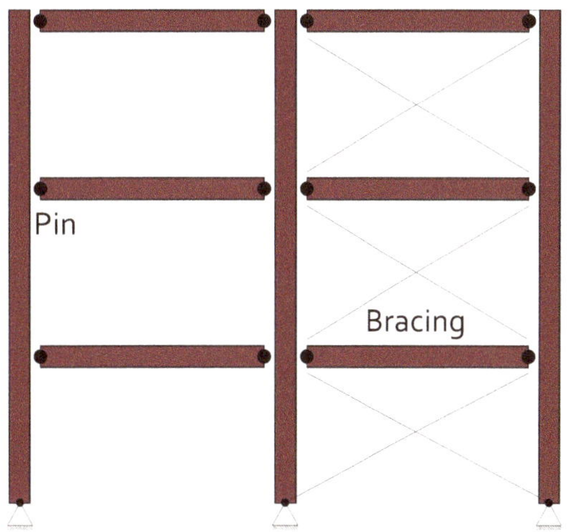

Figure 2.5.6 Braced Frame Beams with Pinned Ends

Multi-storey Cross Braced Frame

A very common form of construction for buildings is a framework: stick elements comprising columns and beams, as illustrated in **figure 2.5.6.**

The most straightforward framing system is referred to in structural engineering as a *braced frame*. This comprises beams fixed to columns with pinned connections.

All tall structures supported off posts or walls need to be kept stable. The destabilising forces of nature are provided by wind, water pressure (sub-sea structures), soil pressure (retaining structures), and gravity (*out of plum or out of true construction*).

The most direct method of restraining or bracing a structure to prevent excessive sideways sway is to cross-brace between beams and columns using flat plates, wires, tubes, and other materials.
The 'X' arrangement ensures that, whichever way the lateral force may act, the frame is restrained by means of cross-braced ties acting in tension or being *stretched*. This is an illustration of *built-in redundancy*.

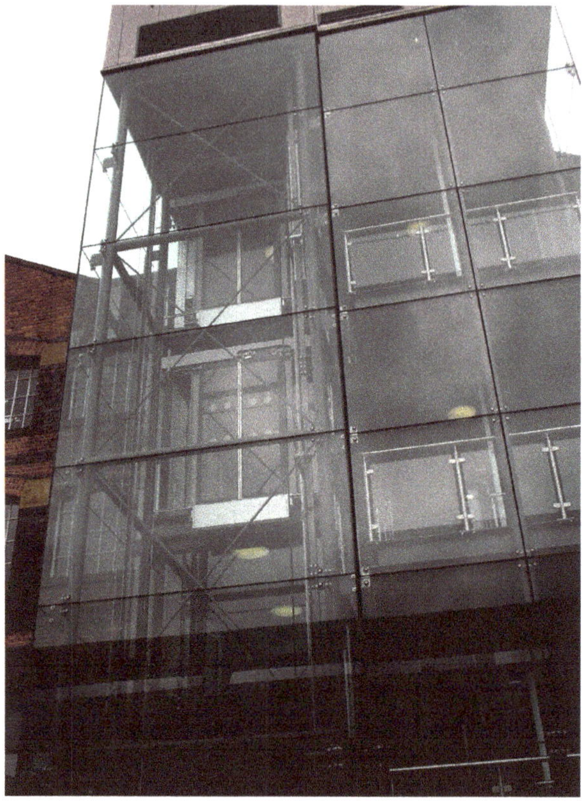

Figure 2.5.7 Braced Frame Example

Aside

Single Strut Bracing

The bracing could be a single sloping strut, which would work as effectively, but the struts would have to resist the reversal load in compression.

Compression struts (as described in later chapters) are subject to buckling. Tension elements have no such concerns and are generally lighter and more efficient in this respect.

Figure 2.5.7 below shows a cross-braced stair tower.

Aircraft are designed with care and attention to structural efficiency and minimisation of weight.

Figure 2.5.8 shows cross-braced wires between the wings and outer struts of a biplane. Although redundant ties are very efficient in restraining lateral distortion. Since the wings are cantilevers, the bracing forms a trussed frame which minimises the cantilever bending moments in the wings. Trusses are reviewed in chapter 2.6.

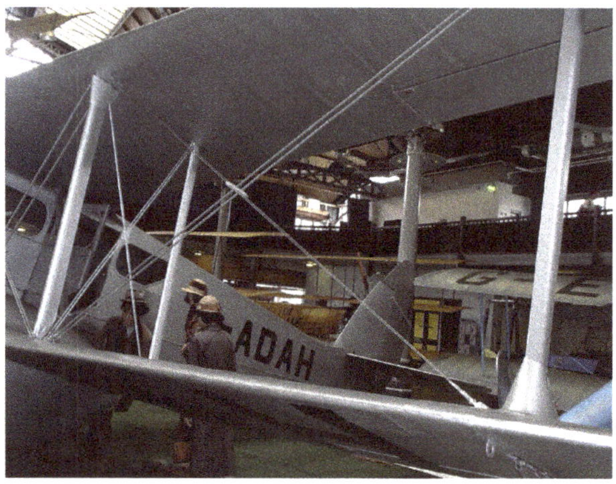

Figure 2.5.8 Biplane With Wire Cross Bracing

Sway or Moment Frames

Sometimes, architectural demands are not easily met with cross-bracing or strut bracing. A building may have a heavily glazed façade with no solid walls behind which the bracing may be hidden. Furthermore, it may have an open-plan interior, with no desire on the part of the client or architect to allow any visible bracing internally.

In this instance, the designer may propose a fixed frame structure without bracing. The restraint provided against lateral loads relies on the columns and beams to absorb the lateral applied forces, this time by allowing bending in the connections.

The foundation supports could be designed to be fixed or pinned (as shown in **figure 2.5.9**).

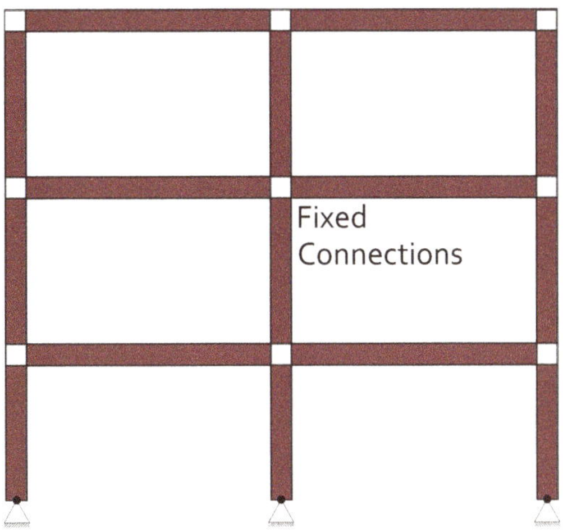

Figure 2.5.9 Multi-storey Fixed Frame

Reflecting on the three-element frames described earlier in this chapter, it should be noted that a fixed frame structure will not be as rigid, with like-for-like sizes assigned to the beams and columns, as a braced frame.

Any releases (such as using pinned supports) will further decrease the overall rigidity or stiffness of the frame. For the engineer, the fixed solution offers architectural benefits but at the likely cost of increased weight of structural elements and more complex and heavy foundations if resorting to fixed connections at the base.

Aside

Fixed Frame Design

More Time, Effort, and Fabrication.....More care and attention is needed with respect to the design of connections between beams and columns (no structural connection ever achieves perfect rigidity) and controlling the frame deflection to acceptable limits. As a rule of thumb, the sway is generally limited to 1/300 to 1/500 of the storey height; that is for the net sway per floor.

Fixed frames offer elegance but at a cost in terms of weight and engineering design time. In steel structures, thicker plates, more bolts, and heavier sections to limit sway deflections may be uneconomical if a braced frame option is possible with some compromise in the architecture.

Wind Frame Analysis

Before the advent of personal computers,engineers resorted to shortcuts which were good approximations of real behaviour. Wind frame analysis was a commonly used method of designing frames up. Fixity in the connections is limited to resisting the wind loading only.

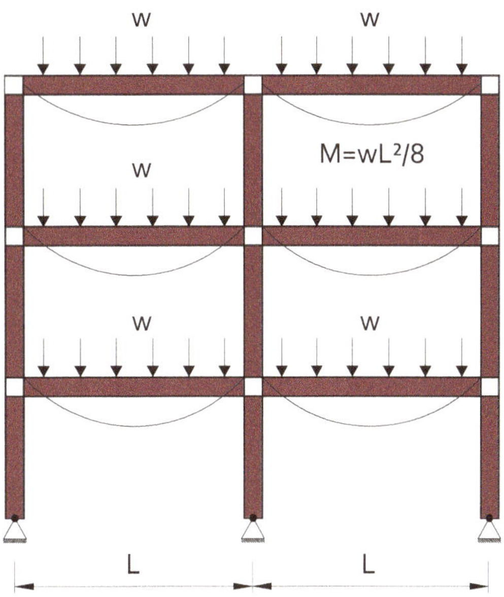

Figure 2.5.10 Wind Frame Floor Loading Moments

Heavily fabricated or complicated fixed connections at the columns are avoided by omitting design fixity for dead and live loading on the beams. This method, deriving from an era that sought short-cuts to design computation, is very effective. It is also useful for approximate sizing and for testing computer models.

Initially, beams are sized as simply supported (see **figure 2.5.10**). In this way, the bending moments and deflections are overestimated since the column connections are fixed.

The second stage has to apply the wind loads or notional horizontal loading to the frame, as illustrated in **figure 2.5.11.**

At this stage, the column/beam connections are assumed to be fully fixed, with virtual pins assumed midspan in the beams and at half floor height in the columns (except for the ground floor columns).

A quick approximation of the connection bending moments is shown in **figure 2.5.11**. In general, the relative magnitude of the wind moments is such that a reasonably economical connection may be designed as opposed to using dead and live loading.

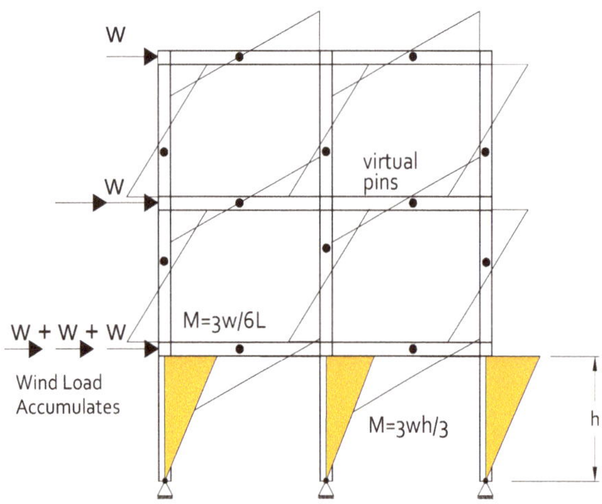

Figure 2.5.11 Wind Frame –Wind Loading Moments

Notes With Respect to Virtual Pins

1. Bending moments from dead and live load are ignored. The design of connections is, therefore, limited to *wind load only*.
2. Wind load at the top of the frame is overestimated (it is closer to W/2 based on half-height). This may be reduced if desired.
3. Deflection check that follows is an approximation for first-stage verification. It ignores element stiffnesses and any base fixity.

Example of Wind Frame Deflection Assessment

The deflection for the bottom storey can be estimated by applying the rules of guided beam deflection or cantilever with tip bending moment (deflection=mh2/2EI, where m=wh).

Recapping the frame will allow some rotation at the joists. The estimated deflection is likely to be greater by approximately 30%.

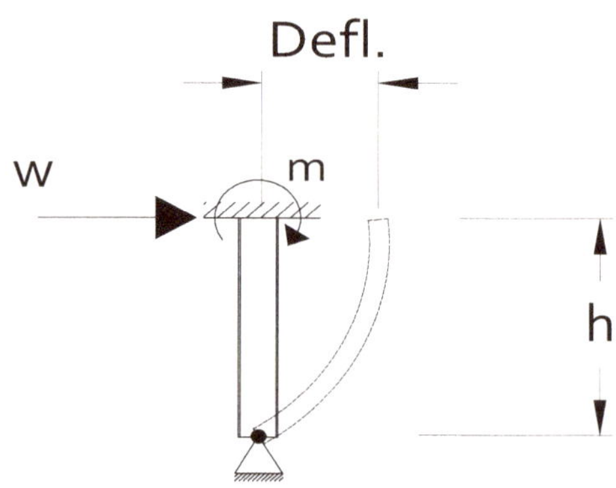

Cantilever With End Moment deflection

$$\text{Deflection} = \frac{Mh^2}{2EI}$$

$$= \frac{wh^3}{2EI}$$

Figure 2.5.12 Wind Frame Base Moments

The upper floors are subject to diminishing wind loads between floors. The deflection at each may be roughly estimated, in the example, by proportioning (pro-rata) the calculated deflection according to the wind load in the floor above that considered.

Using the column stiffness (EI value) from the model, the equation in **figure 2.5.12** was applied

Comparison of Deflections

The estimated deflections were compared with the output from a three-storey frame modelled on a 2D computer model using the same columns as above.

The deflection from the model at each level was:

1st Floor:	14.8mm
2nd Floor:	21.0mm
3rd Floor:	24.0mm

The cumulative deflection is shown in the estimate following, based on the assumption that the deflections reduce in proportion to the cumulative wind load at each floor.

In summary, the total estimated deflection was in the order of 13% of the computer model. This is a reasonable approximation.

Recap (Wind Frames)

Calculate Total Deflections at Top

Total (estimated) Deflection
= 11.6 (GF) + 11.6/2 (1st) + 11.6/3 (2nd)

Top deflection = **21.2 mm**

....compared with 24mm (13% Difference).

In 4 steps of calculation, namely:

1. Bending Moment (simply supported)
2. Midspan beam Deflection (simply supported)
3. Column loading (cumulative beam shear addition from roof)
4. Column sway

The wind frame method is a handy method that avoids a large number of computations.

Goal Post Frame

Goal post frames are aptly named double posts with single beam frames. In smaller and domestic scale construction projects the post and beam construction is used commonly. One particular use is for restraining gable walls of houses around large window or door openings, thereby stiffening up the masonry or timber frame or similar.

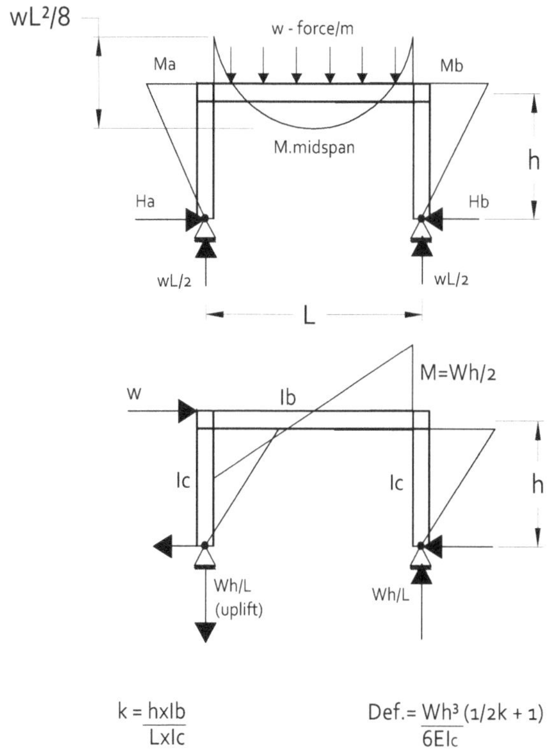

Figure 2.5.13 Simple Goal Post Frame

They may be simply modelled – as for the frame in **figure 2.5.13**, which has pinned bases. A different approach takes advantage of the relatively simple empirical equations which may be obtained for single-storey goal post frames. These equations are summarised in **figures 2.5.13** and **2.5.14**.

It is recommended to make use of this simple condition for checking frames. The column/beam connection should be designed for wind, dead, and live loads combined as necessary by *superimposition*.

If the bases are fixed, similar equations are available. Fixing the base of a goal post frame is very effective in reducing the overall deflection. The deflection is roughly halved: the columns are like guided beams with both ends restrained from rotating.

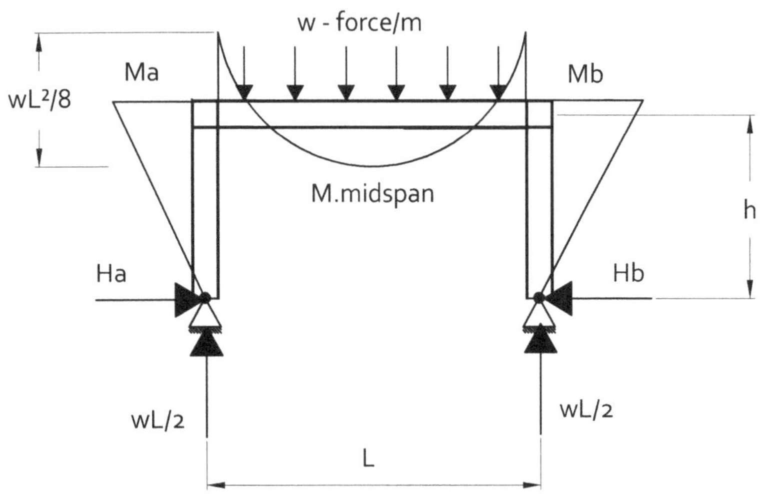

$$k = \frac{h \times Ib}{L \times Ic}$$

$$n = 2k + 3$$

$$\text{Def. mid.} = \frac{5wL^4}{384EIb} - 0.80 \times 2\,\frac{MbL^2}{16EIb}$$

$$Mb = Mc = \frac{-wL^2}{4N}$$

$$\text{M.midspan} = \frac{wL^2}{8} - Mb$$

$$Ha = Hb = \frac{Mb}{h}$$

Figure 2.5.14 Simple Goal Post Frame Bending and Shear

Making Use of Partial Fixity:

It is possible in the computer models to assign partially fixed bases. In this way, a base fixing and foundation with significantly reduced moments may be designed. It will be found that adding a partially fixed base significantly reduces deflections. This can easily be verified on a computer.

Assessing the fixity usually requires a trial and error approach, testing varying values for the column bases since assessing the actual stiffness is almost impossible with the varying factors of baseplate stiffness and ground stiffness.

For lateral deflections the frame sway should be limited to height/300 for a clad structure to height/500 for a frame bracing a masonry structure. In other words, for a storey height of 2.5m, deflections should be limited to approximately 5-10mm. Assuming a partially fixed base can assist with designing an economic frame.

Wind Frame Example

The example below introduces calculations in SI units (Force in kN and dimensions in m). From this point, examples in the following chapters will be using the same units.

Example ref. Figure 2.5.13

Wind Frame Parameters:

H (height) post, h	= 2.5m
Span (beam), L	= 5m
Wind Load at eaves, W	= 10kN
Vertical Loading on Beam, w	= 20kN/m

Post/Column (UK Section 203x133x25UB)
Moment of inertia, Ic = 2400 cm^4

Beam (UK Section 254x146x31UB)
Moment of Inertia, Ib = 4500 cm^4

Calculations will follow to ascertain bending moments and deflections, based on **figures 2.5.13** and **2.5.14**.

Following from the previous example:

Calculate Bending Moments in Post From Wind Loading

Mc (W) = Wh/2 = 10x2.5m/2 = 12.5kNm

Calculate Bending Moments in Beam from Beam UDL Loading

Calculate Constant, k (figure 2.5.13/2.5.14)

K = (hxIb)/(LxIc) = $2.5 \times 45 \times 10^6/(5 \times 24 \times 10^6)$ = **0.94**

Calculate Constant, n (figure 2.5.14)

n = 2k+3 = 2x0.94 + 3 = **4.88**

Calculate Beam End Moment, Ma (=Mb)

Ma = $wL^2/4n$ = $20kN/m \times 5^2/(4 \times 4.88)$ = **25.6kNm**

Calculate Beam Midspan Moment, M.midspan

M.midspan = $wl^2/8$ - Mb
= 20x52/8 - 25.6 = **36.9kNm**

Calculate Wind Loading Sway Deflection (ref. figure 2.5.13)

E (steel) = 210×10^3 mPa (N/mm^2)
Deflection (sway) = $(WL^3/(6EIc)) \times (1/2k+1)$
=$\{10 \times 10^3 N \times (2.500 \times 10^3)^3/6/210 \times 10^3/2400 \times 10^4\}$ x (1/(2x0.94)+1) = 5.16 x (0.53+1) = **7.9mm**

Calculate Beam Midspan Deflection

This calculation is an approximation, but suffices for the purpose. In general the wind sway deflection is the more critical for design of goal post frames.

Deflection Midspan (ref. Figure 2.5.14)

= $5wL^4/384EIb - 0.8 \times 2\ MbL^2/16EI$

= $5 \times 20N/mm \times 5000^4/384/210 \times 10^3/4500 \times 10^4 - 0.8 \times 2 \times 25.6 \times 10^6 \times 5000^2/16/210 \times 10^3/4500 \times 10^4$

= 17.6mm - 6.8mm = **10.8mm**

Aside

Buildings as Braced Frames

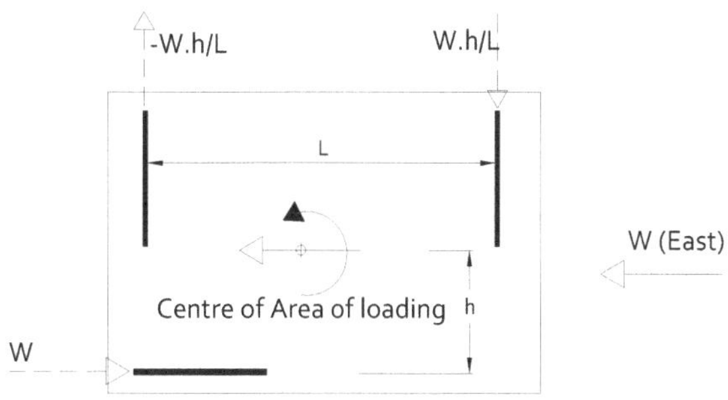

Figure 2.5.15 East Wind on Braced Frame

Braced frames are located as close as architecturally possible to the perimeter in a manner such that they will resist the lateral forces of wind and possible 'out of balance' vertical loads, which need lateral restraint to balance the system. A very good reason for selecting a braced frame is that it eliminates the need for the fabrication of numerous fixed connections. Column design is also simpler; columns may have an effective length of 1.0 or less.

A simplified review of how frames interact in a building a simple arrangement is shown in figures 2.5.15 and 2.5.16. A rectangular plan building in steel has cross-braced frames placed adjacent to 3 sides. This is the least number of frames that will resist wind loading that may arise from any direction. In the case of an East (or West) wind, the only braced frame is offset from the centre of loading by a distance 'h', as shown in figure 2.5.15. Twist is resisted by the North/South frames.

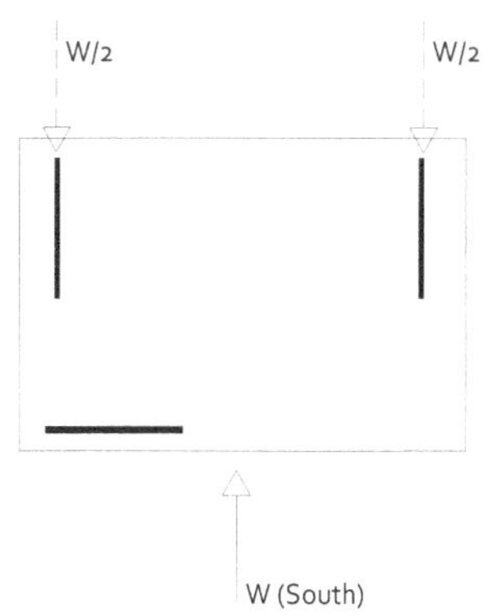

Figure 2.5.16 South Wind On Braced Frame

Chapter 2.5
Quick Note Summary

1. Framed structures are generally designed as braced or unbraced.

2. The most simple and economical frame to design is the braced frame in terms of element sizes and fabrication techniques.

3. Unbraced structures use connection fixity between beams and columns. Partial rotation occurs at supports, which is difficult to quantify by hand calculation.

4. The wind frame method is a good way to estimate the sizes of structural elements and estimated deflections with simple hand calculation checks.

5. Goal post frames are the most simple form of wind frames; they are used widely in domestic structures around wide openings to stabilise walls.

6. Braced frames are an effective and economical method of stabilising buildings. Consideration needs to be given to the distribution of forces that are imparted into the braced frames. Locating braced frames (or shear walls in the case of concrete frames) within the building footprint is an important consideration to prevent building twists.

7. A good rule of thumb for the design of columns in a frame is to use an effective length factor of 1 (or less) for a braced frame and 1 to 1.5 for a fixed or sway frame. This is alluded to in **chapter 2.7** (Euler buckling).

Chapter 2.6

2.6 Trusses and Vierendeel Structures

Trusses are probably best known as long-span, deep frameworks, which are used for bridging large voids in buildings or across wide bays and valleys. Other widely used applications are roof supports – as timber, iron, or steel frameworks – from medieval times until today. Vierendeel girders, named after the inventor - a Belgian engineer - date from the twentieth century. They ultilise fixed connections. They are included with trusses since the two are most often substituted depending on circumstances. The chapter summarises the use of virtual cut sections for speedy resolution of forces; aspects of stiffness (by the use of beam analogy); trusses as deep beams; trusses as composite beam/ pin jointed applications (king and queen post trusses); variations of roof trusses and vierendeel girders using approximate design methods.

Trusses

Trusses have been used in multiple forms throughout the last 200 years. They have mainly been built in timber, wrought iron, and steel, with new forms developed in the 19th century. In the early development of the railways in the United States, with an abundance of timber, railway bridges were built quickly and effectively in timber, but this later gave way to more use of wrought iron and steel.

Timber was extensively used as a material to build roof trusses, sometimes with steel rods or angles used for the tension elements. An advantage of the truss form in timber is that large-span structures can be built up, which are not limited by the available lengths of timber.

The nineteenth century was a period of the invention of trusses. A number of types are summarised in the following list. In this instance the internet can provide in minutes a lot of images and descriptions of trusses and their history.

Railways/Transport

- Warren Truss (see figure 2.6.1)
- Fink Truss
- Bowstring Truss
- Pratt Truss
- Lattice Girder

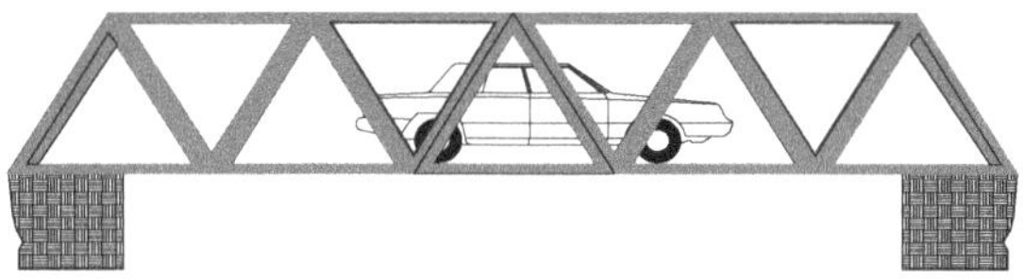

Figure 2.6.1 Warren Truss

Roofs (pedestrian bridges)

- Polonceau Truss (roofs)
- Belgian truss
- Belfast truss
- King Post
- Queen Post
- Howe Truss

Trussed Beams (Floors)

- Modified Warren truss
- Lattice Girder

Trusses and 'Trussed Beams' the Difference?

A simple distinction will be made between a truss and a *trussed beam*. Trusses are medium to deep beams which are efficient forms of structure deriving structural rigidity and strength from internal stick components connected together with simple pin joints. They are of practical use where their high depth-to-span ratio is possible without compromising useful space, for instance, in a bridge where they flank the road or railway supporting deck.

Trusses are employed frequently in buildings. If the full floor depth is available, they may be used, for example, as transfer structures to carry the upper floors in a multi-storey building. In **figure 2.6.2** the upper floors of a building are offices with regularly spaced columns. A deep truss on the first floor supports the upper-storey columns over an open retail space on the ground floor.

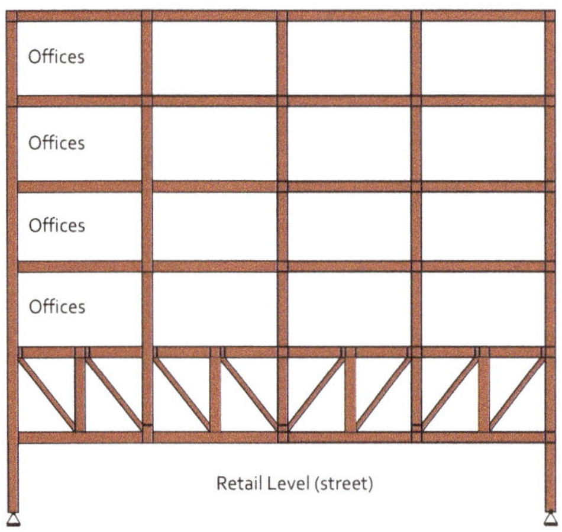

Figure 2.6.2 Multi Storey Transfer Truss

Trussed beams are generally used within floors or bridge decks where their depth-to-span ratio is shallow (span/depth 10-20). The beams in generic form comprise upper and lower flanges – for instance, timber joists – with the web formed in lightweight lattice material – often pressed, thin gauge steel.

This type of structure is a hybrid between a beam and a truss: the bending stiffness of the top and bottom flanges works along with the truss action, resisting the shear, of the internal components.

Service ducts may be run within the web zone.

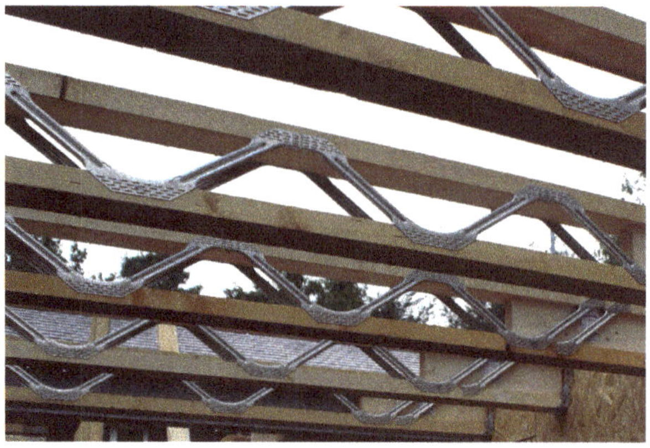

Figure 2.6.3 Web Truss Joists

Internal Forces

Trusses carry loads with the internal members loaded in a purely axial direction (either tension or compression) except for hybrid types such as the kingpost and web joist trusses. The overall distribution of forces conforms to equilibrium just as a simply supported beam. Like beams, trusses may be made to work as continuous structures or cantilever structures.

Examination of a Warren Truss

Figure 2.6.4 shows two forms of warren truss. The bottom truss is identical to the top truss but simply inverted. Importantly, the right-hand side (RHS) support is allowed to slide sideways. This ensures that the truss is a self-contained structure that does not transmit thermal or other lateral forces onto its supporting posts or abutments.

The structural action of a truss is analogous to that of a beam: the flexural action causes tensile forces in the bottom chord (the bottom continuous element) and compressive forces in the top chord for a simply supported condition.

The diagonal struts between the top and bottom chord carry the shear within the truss. This is an important property: it allows the minimising of material within with relatively lightweight internal elements (see **figure 2.6.3** web truss joists where the internal pressed steel is lightweight).

It may be remembered that the internal elements support the forces only in tension and compression. There is no shear force or bending moment to be considered within the internal elements. An internal axial force distribution is shown as **'T'** for tension and **'C'** for compression in **figure 2.6.4**.

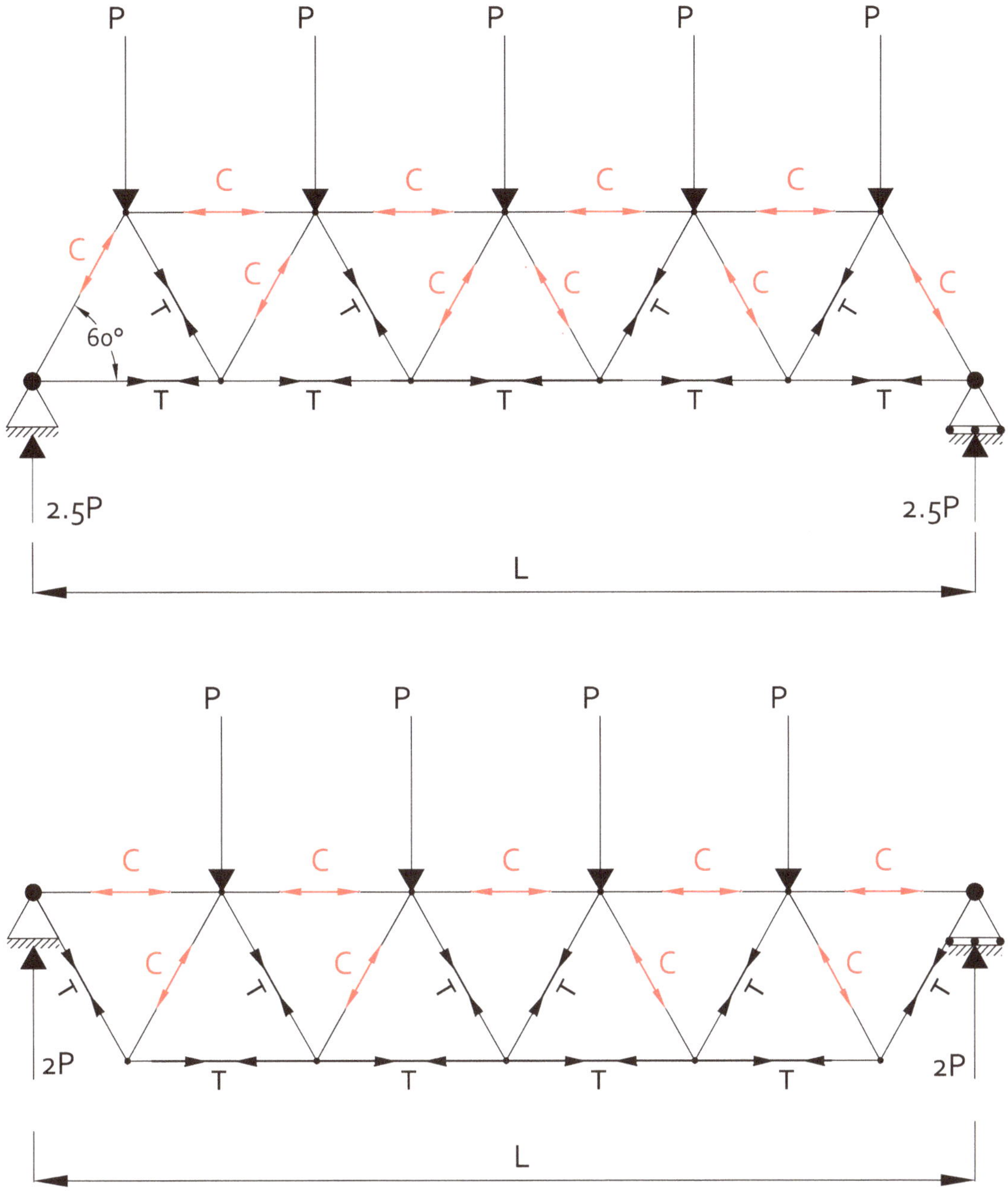

Figure 2.6.4 Warren Truss Two Options

Aside Inverted Truss

It is to be noted that the inverted truss is identical in its load-carrying capacity, except that the compression and tension members are now swapped, excluding the top and bottom chords, within the diagonal struts. Note: compression in the top chord and tension in the bottom chord- same for both examples.

For the inverted truss, the first internal 'shear' carrying strut is in tension rather than compression. This orientation may possibly be constructed of lighter sections for reasons that are described in the next part.

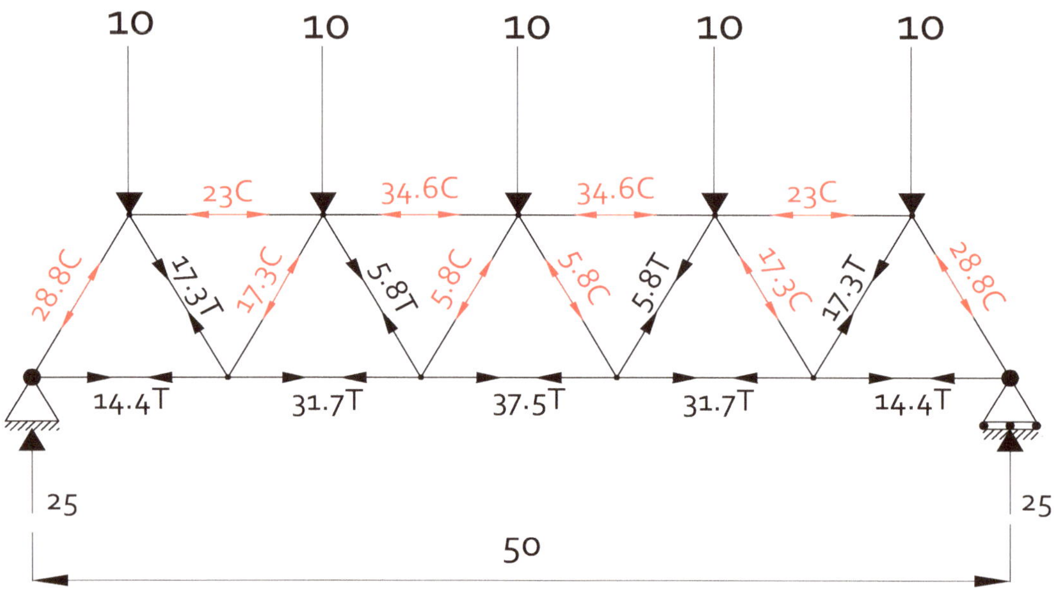

Figure 2.6.5 Warren Truss Internal Forces

Numerical Study of Internal Forces by Equilibrium at a Joint

The warren truss in **figure 2.6.5** was based on a hand analysis that was undertaken in SI units. The actual units have been omitted to clarify the methods involved. The internal forces may be found by a set of sequential equations.

Note:

- **23C** denotes a compression force, value 23
- **31.7T** denotes a tension force, value 31.7.

To achieve statical equilibrium, which is important for solving the internal truss forces:

- External forces balance External forces.
- Internal forces balance Internal Forces (each joint balances).

A virtual cut around joints (or across the truss) needs virtual external forces to balance the cut members' internal forces.

Calculating Internal forces

This method is error-prone; a methodical procedure in which it is easy to lose track of the overall forces.

Short Cut to Designing a Truss

Since trusses are generally rationalised by making the top and bottom chords the same section size, together with the internal struts and ties, shortcuts may be taken.

Resolve for maximum strut forces at the support: in **figure 2.6.5** the first internal struts are the highest loaded (this is the location of maximum shear). Resolving vertically for the calculated reaction, only a single equation is needed to find the internal compression in this strut (**figure 2.6.6**).

For simplicity, the section may be selected, and the other struts and ties given the same section size.

Resolve for maximum bending moment location for the maximum top and bottom chord forces: the midspan bending moment is divided by the distance between the centre of area of the top and bottom chord (**figure 2.6.7**), giving the internal force in tension (bottom chord) and compression (top chord). The chords may now be sized by taking the critical compression load and comparing it with the allowable compression in a section between the lateral restraints (direction into the page).

Approximation of End Strut Force (C) - ref. figure 2.6.6:

RA = C.Cos30°....therefore

C= 25/0.866 = 28.8.

Calculation of chord force (CF) - ref **figure 2.6.7**

Calculate Moment () - assuming simply supported with UDL....

M= (10/10)x50²/8 = 312.5

Hence CF......(d=depth of truss)........

CF = M/d = 312.5/8.66 = **36.0**

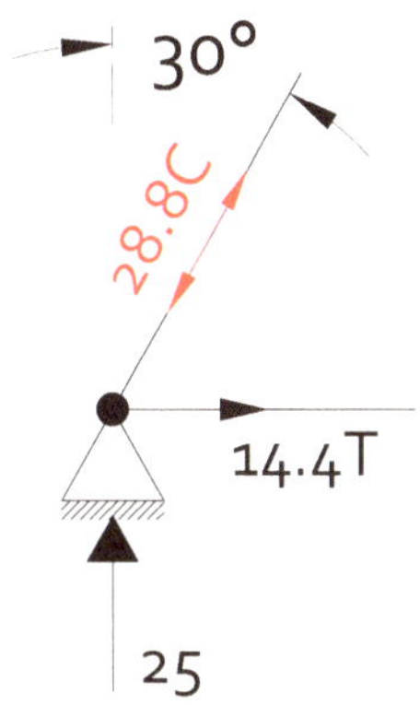

Figure 2.6.6 Resolution of Forces at Support

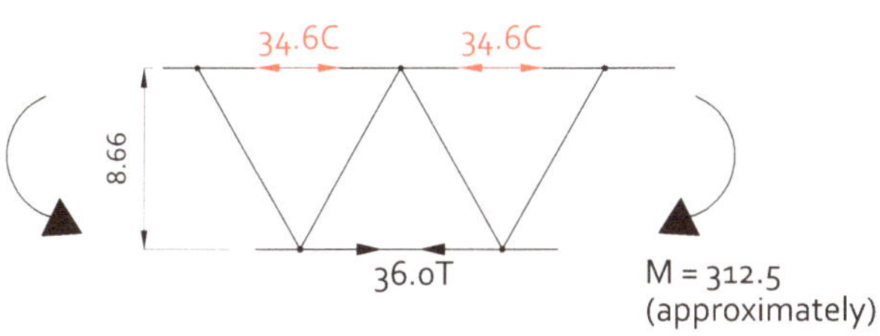

Figure 2.6.7 Chord Sizing

Finding Internal Forces by 'Virtual Cutting'

An elegant method of finding the internal forces is to make a virtual cut through the truss. A central cut was made below.

The internal forces of the cut members then become *virtual external forces* to achieve equilibrium. See **figure 2.6.8**.

Figure 2.6.8 and **figure 2.6.9** show how the chord forces may be derived. The top chord forces are resolved horizontally from **figure 2.6.8**.

Based on the truss in **figure 2.6.4**, the top and bottom chord horizontal loading balances the external forces to the LHS of the truss. These are the virtual external forces that would be needed to hold up the truss in statical equilibrium.

The central moment is found by statical equilibrium. This was found more accurately by using point loading (the previous example calculated the central moment using a virtual UDL), but a virtual UDL method suffices for approximate purposes.

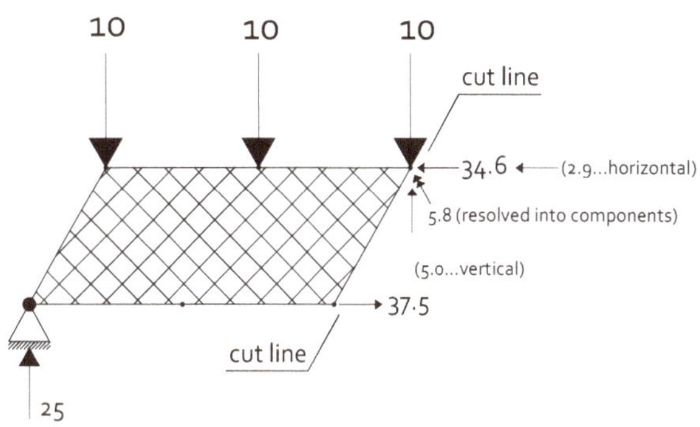

Figure 2.6.8 Virtual Cut Section Centre

Finding Chord Forces

Top and bottom chords balance the moment: referring to **figure 2.6.9** it can be found, by resolving the diagonal cut strut into horizontal and vertical components, that the total force in the top chord (37.5) balances the force in the bottom chord.

Sum of horizontal forces is zero: TThere is no horizontal reaction in sliding supports (see **figure 2.6.8**). The inequality of top and bottom chord loads is caused by the existence of a force in the central struts.

Balance of vertical forces: : The vertical component in the diagonal (**5**) is half the external force at that point – subtracting this from the total applied loads (**30**) equals the reaction (**25**). This component carried by the central strut. See **figure 2.6.8**.

It is possible to find all of the internal forces of the cut elements with three equations: resolving vertical forces, horizontal forces and moments.

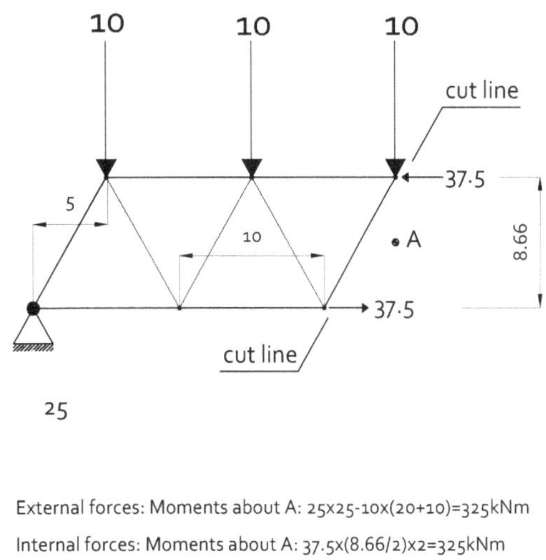

Figure 2.6.9 Resolution of Chord Forces

Taking a beam analogy, the centre of the truss is the position of maximum bending in the structural system. The top and bottom chords are resisting the bending as a couple, with the lever arm equal to the depth (**8.66**).

The centre is the location of zero shear for a quasi-UDL– or uniform loading.

Maximum Strut/Tie Load

Finding the strut loading at the support is by taking a cut at the end of the truss, the same principle as a central cut. Using joint resolution at the support resolves the first internal strut force of **28.8** *(compression)*, ref. **figure 2.6.6**.

Deflections

Deflections were traditionally found by a method of *virtual work* using a summation of strains within the internal members. With computer analysis now virtually universal for trusses, this is seldom used.

An alternative approximate method that equated the truss stiffness to an equivalent beam is discussed in the following section.

In earlier times, this was also undertaken with graphical methods, which were still taught at universities in the late 20th century.

Simplified methods are possible utilising beam analogy.

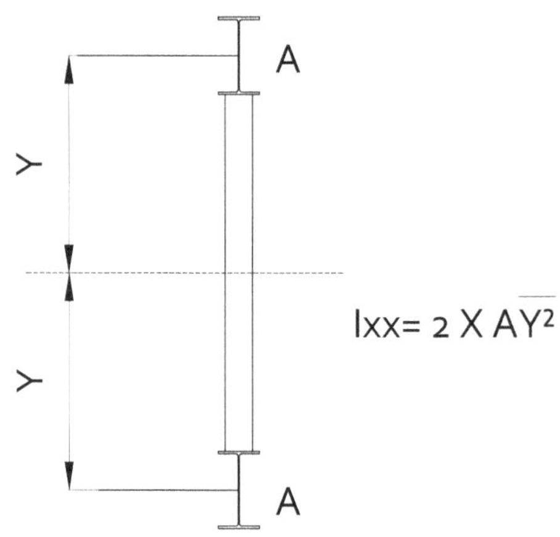

Figure 2.6.10 Virtual Moment of Inertia

Short cuts – Finding Deflections

Following the beam analogy, an approximate method of finding the deflection of a truss is to calculate an idealised I (moment of inertia).

Ignoring the stiffening effect of the diagonal struts, the Inertia of the truss in **figure 2.6.5** may be approximated to:

$I = \Sigma 2Ay^2$

Where y is the distance from the centroid to the centre of area of the top or bottom chord.

If the top and bottom chords are the same section (as noted previously) the approximate moment of inertia may be calculated as shown.

Making a further simplification of the truss, the point loads are approximated to a UDL. This is simply by summing all of the forces and dividing by the span (or dividing the point loads by the distance between them).

See **figures 2.2.9** to **2.2.12** in the section on simply supported beams. Finally, with the idealised beam and UDL, the deflection calculation is simple.

The deflection calculation for idealised UDL and inertia (I) is:

Midspan deflection = $5Wl^4/384EI$

....where I is calculated as per **figure 2.6.10**

A calculation was carried out in SI units for the above with A=5100mm².

Using the simply supported formula and comparing with a truss analysis using the same elements, the results are summarised below:

Truss Analysis:	2.8mm
$5Wl^4/384EI$ estimate:	2.0mm

The answer has around 40% accuracy, which is sufficient to give an estimate for the purposes of carrying out a computer analysis. But where is the missing 0.8mm?

Shear Deflection Contribution

The answer to the missing 0.8mm lies in the deep beam analogy. Trusses are generally deep in comparison with their span. The example in the previous sections has a depth-to-span ratio of 1 in 6. According to the shear deflection example (ref. **figure 2.2.6**) the contribution of shear is significant.

Given that a truss is *not quite like a beam*, an estimate of the shear contribution may be found by assuming the following:

Effective shear plate area is equivalent to: Strut area (5100) Shear deflection formula for simply supported beam with UDL: Shear deflection = $wL^2\alpha/8AG$

α is a form factor depending on the section type – 2 for an I-beam section 6/5 for a rectangular section. This analogy is assumed to be a rectangular section.

G is the shear modulus, which has not yet been examined but which may be found in material tables. G is to shear equations and torsion formulae, as E is to flexural (or bending) formulae.

A is the area of 2 diagonal struts.

Resultant shear deflection estimate:
(truss ref. **figure 2.6.5**) = **0.94mm**

Resultant Shear and Bending Deformation

Combining the estimate shear and bending

Bending Deformation	= 2.0mm
Shear Deformation	= 0.94mm
Combination	= 2.94mm

This is within 5% of the analysed figure of **2.80mm**.

Obtaining truss deflections by this method is not a rigorous process, but it provides useful approximations when checking computer output.

Shear deformation may be significant in trusses.

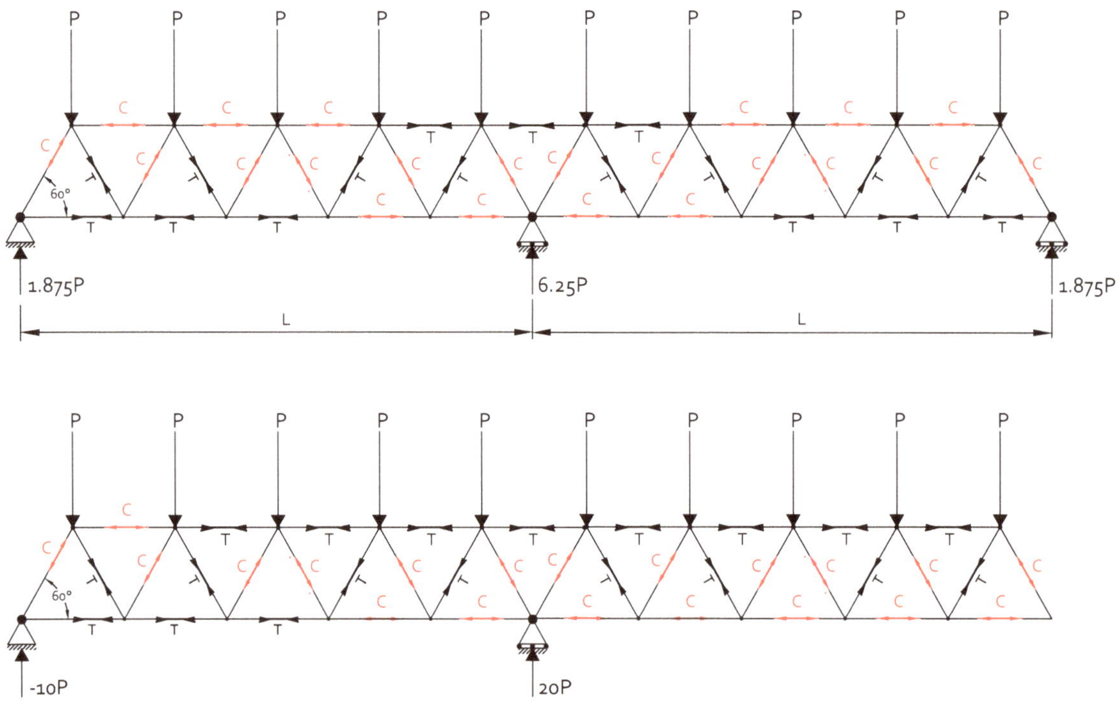

Figure 2.6.11 Warren Truss - Continous and Cantilevered

Trusses as Cantilevers and Continuous Beams

Continuing with the beam analogy, trusses may be employed as continuous beams or cantilevers.

The most famous cantilever truss bridge probably is the Forth of Firth Bridge, shown in **figure 2.6.12**.

Three *cantilever towers* support internal spans, which are smaller, simply supported trusses. Note in the photograph above how the compression elements are stout tubes (bottom part of cantilever truss) and the top tension elements are fabricated sections for the cantilever towers. The internal spans are essentially pinned, conforming to the principle of hinge joints illustrated in **figure 2.4.9**.

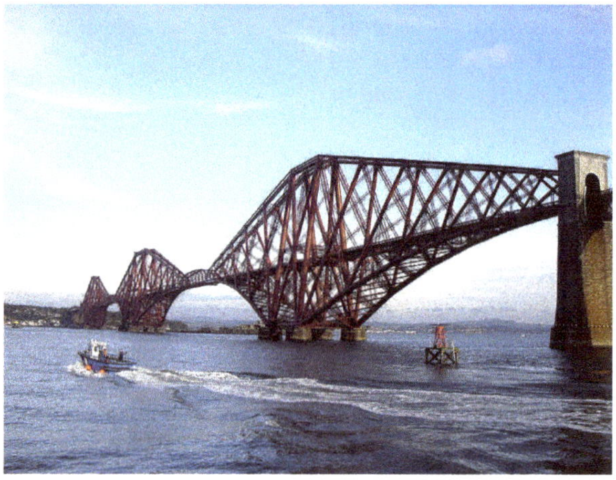

Figure 2.6.12 Forth Bridge - Multispan Truss

Hybrid Trusses (King Post/Queenpost)

A common form of truss in buildings employed as floor beams, or in an inverted form, as roof trusses is the kingpost truss (see **figure 2.6.16**). A structure that may best be described as a beam with stiffening ties.

Figure 2.6.13 shows a kingpost arrangement. An inverted strut sits below the centre of a beam with ties running from the tip of the strut to the support point. The rake angle for this form is customarily 20-30°. This is the slope of the tie bar (Φ ref. **figure 2.6.14**).

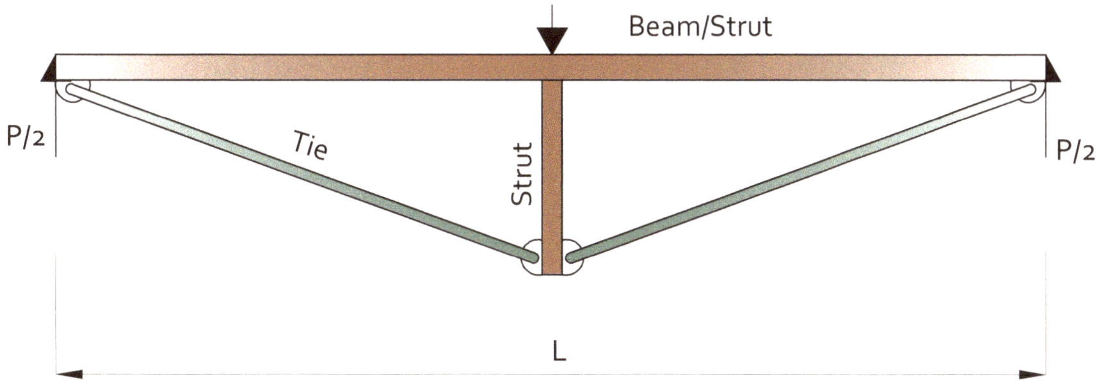

Figure 2.6.13 King Post Truss

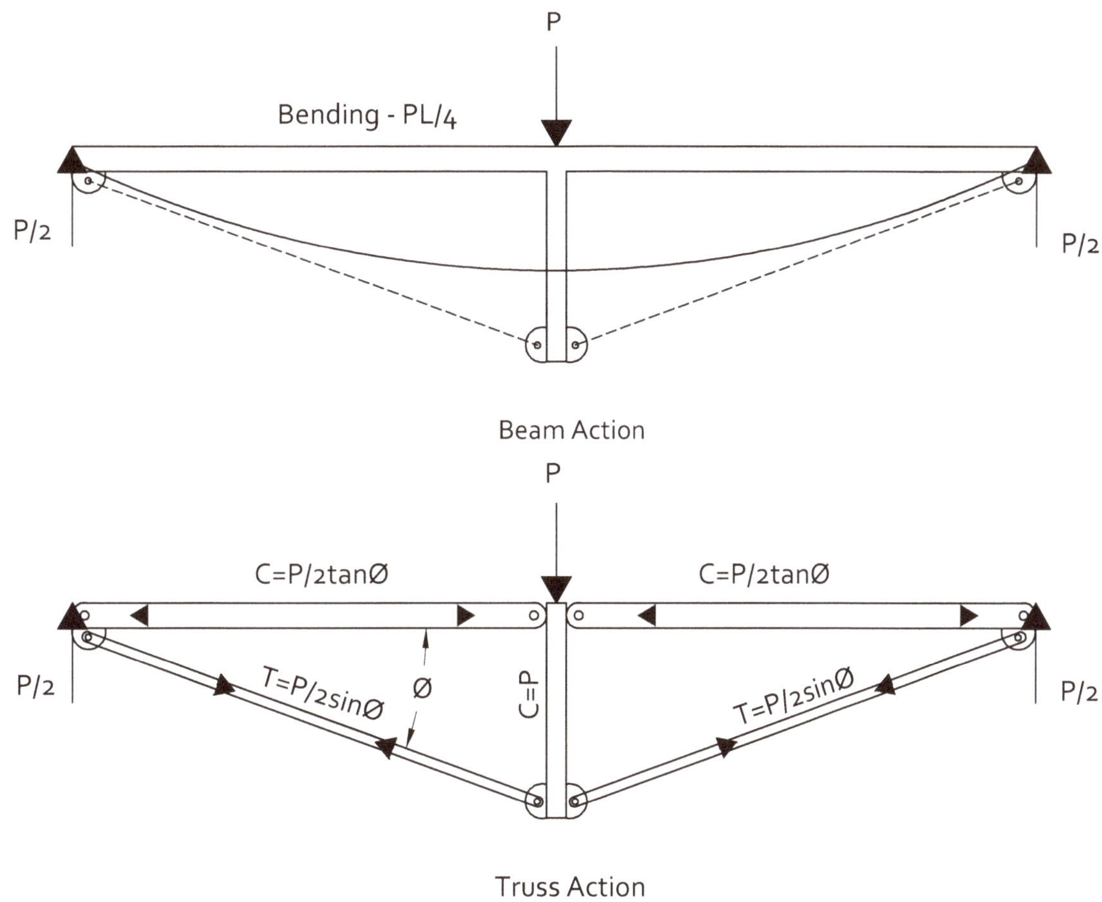

Figure 2.6.14 King Post Truss - Hybrid Action

Hybrid Truss/Beam Action

Figure 2.6.14 shows the two defining structural actions of this hybrid form of structure: that of a beam and that of a truss. Resolution of the internal forces cannot be found by either beam forces or truss action (assuming a hinge formed within the centre of the beam).

The combined action may be found by combining the equations by use of *compatible deflections.* The degree to which it behaves as a truss or beam depends on the relative stiffness of the truss and the beam action.

Kingpost Bridge Example

The truss in **figure 2.6.15** behaves in a manner similar to the kingpost: with stiff top and bottom chords, or *flanges*, the beams act as a hybrid between beam (action of the flanges) and truss (action of the struts).

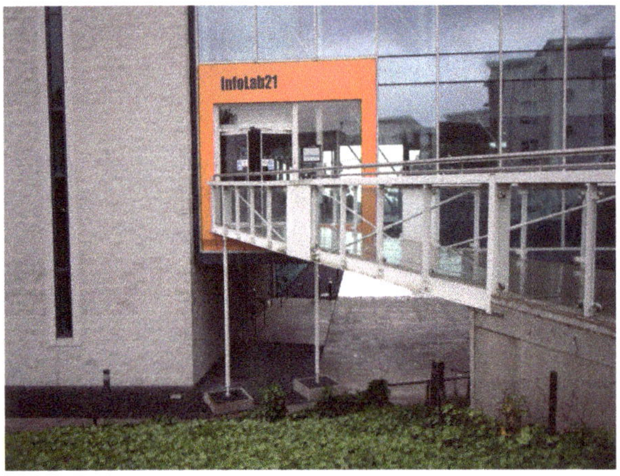

Figure 2.6.15 Infolab Bridge -Kingpost Truss

This footbridge comprises a top handrail, which is the *beam* and tie rods placed within the balustrade.

The top handrail (channel section) and bottom chord (angle section) are not particularly stiff to avoid visual intrusion. The bridge has more trusses than *beamness*.

Because the top chord operates in compression, there is the issue of buckling (see discussion in **section 2.7, page 78** – ref. *Euler*).

The channel also acts as a handrail, which needs lateral stiffness.

By utilising the end posts and central posts as rigid elements, with stiff bases resisting rotation, the top chord was designed to span laterally between these restraints.

The intermediate plate glass restraints could then be sized as slender plates with pinned connections to the top and bottom, thus avoiding a series of heavy posts.

Traditional Forms of Roof Truss

Common to 19th-century industrial buildings in the UK is a form of roof truss known as the kingpost. A 1925 carpentry and joinery book extract is shown in **figure 2.6.16** of this form of truss.

The kingpost roof truss is an inverted form of the truss shown in **figures 2.6.13, 2.6.14,** and **2.6.15**. The principal elements are described below:

- Internal Raking Struts. Two internal struts are generally added. These sit under the rafter supporting beams (*purlins*), preventing excessive sagging from the direct purlin load. These struts act in compression with a tongue and groove or notched fitting, usually sufficient to transfer the load

- Bottom Chord (or Tie Beam).The beam entitled tie beam at the bottom (the bottom chord) is generally a stout section, often made of Baltic pine or a similar robust type of timber. As discussed in the previous section, the bottom chord behaves as a combined beam and tie. It is an elegant, simple form of roof support.

- Top Chord (or Principal Rafter). This element follows the slope of the roof, supporting the roof purlins and ridge beams (which support the rafters). It is in compression, notched into the bottom chord to transfer the load.

- Kingpost. Central vertical post from which the bottom chord hangs, connected by a strap or stirrup with cotter pins or a bolt. This element acts in tension.

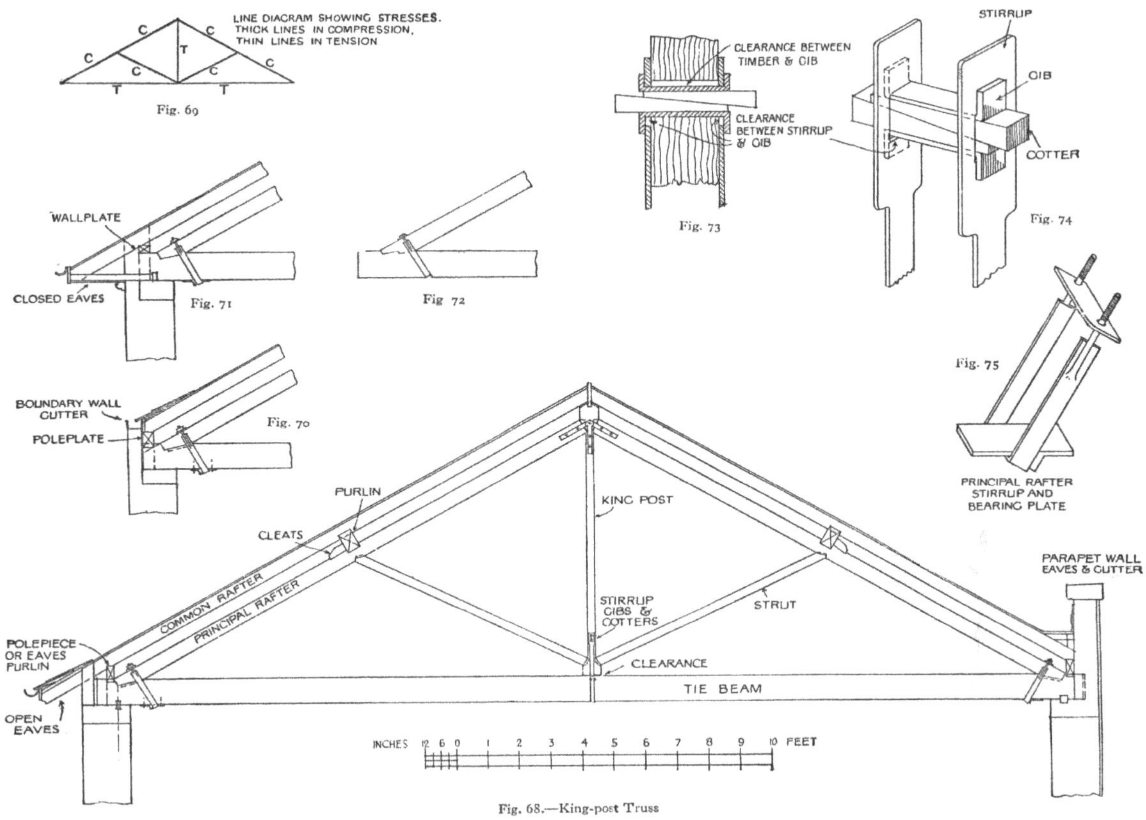

Fig. 68.—King-post Truss

Figure 2.6.16 Traditional King Post Timber Truss

Extract From Carpentry and Joinery (Ednie 1925, reference 3)

Principal Rafter Slip (long term)

Now that many of the 19th-century buildings are approaching 150 years old or more, this form of a truss is often to be found where many of the warehouses have been converted to office buildings or apartment blocks frequently as an architectural feature.

Perhaps it was for good reason that the bottom chord was stout. In many instances, over time, two occurrences tend to weaken the truss action. At eaves with failed or leaking gutters, dampness causes rot in the joint between the bottom chord (tie beam) and the principal rafter.

Often, the compression in the rafter will cause slipping of the joint. Also, the central struts often shrink, causing further bowing.

A cautionary note for modelling existing trusses:

Ensure the bottom chord is a continuous member from support to support without a central pin.

Assessment and Repair of Historic Trusses

To assess the *weakened* truss the top rafters may be removed and the loads applied with the beam in pure bending. This is likely to indicate excessive sagging in line with the observed deflection of the existing truss.

Repair may involve re-propping and jacking up the tie beam and refitting the stirrup.

Repair of the ends of rotted tie beams may necessitate attaching new timber to the end.

Modern methods of attachment commonly involve the use of rods inserted into drilled holes on each side and resin grouted in place.

The new end timber can be notched and a steel shoe fitted if necessary to accept the principal rafter.

Roof Truss Variations

Figure 2.6.17 illustrates a set of roof truss options sketched for a client who was looking for a historic vernacular.

There is an industry in producing traditional timber trusses. An internet search will reveal a wide range of different kinds of trusses.

Selection of options may depend on a number of factors aside from aesthetic concerns. For instance, in a very deep roof, the queen-post truss is often favoured because it allows usable attic space (as long as the middle horizontal strut allows sufficient headroom under).

Steel bars may be used as tension elements in combination with timber or even concrete.

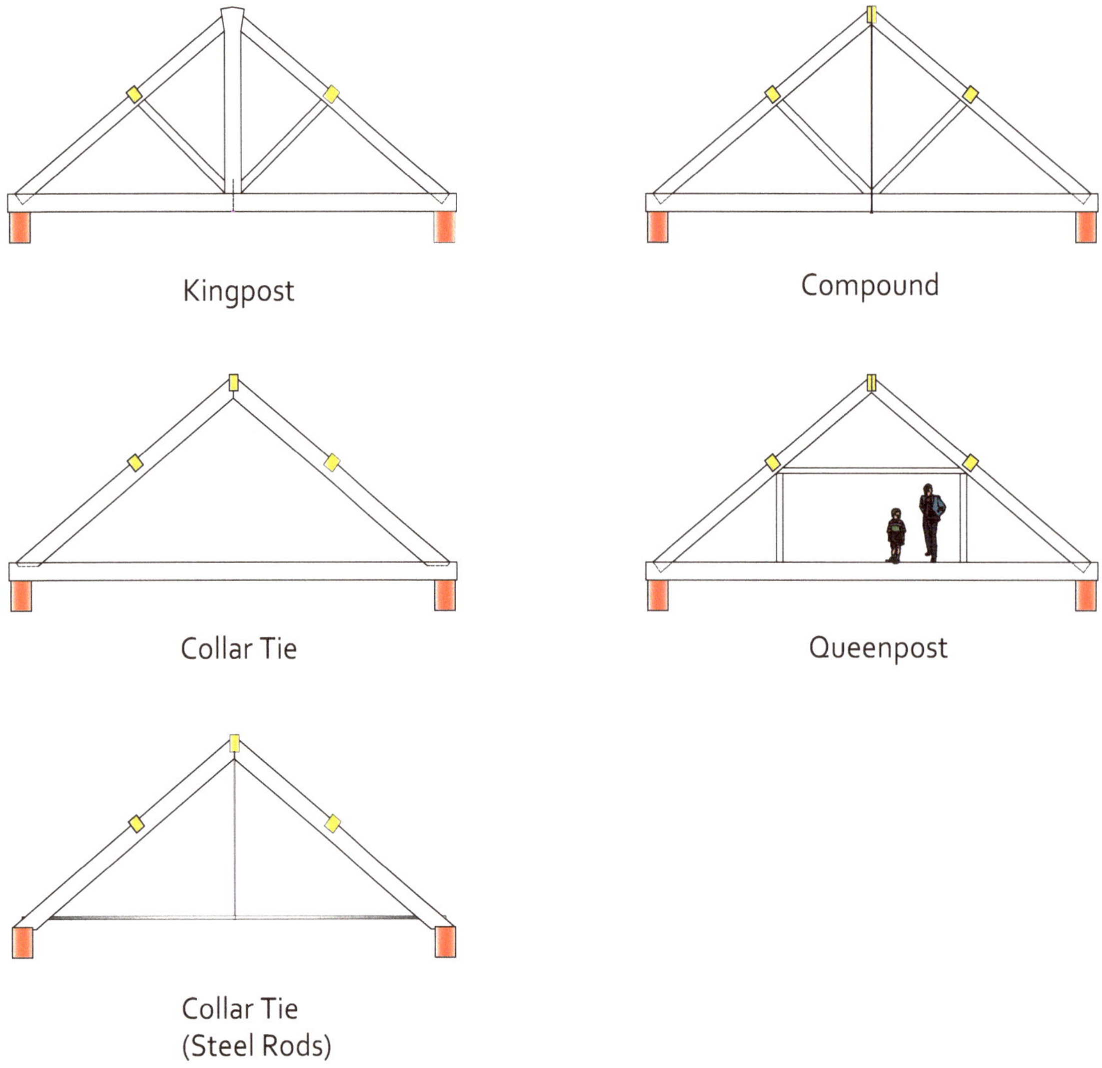

Figure 2.6.17 Roof Truss Variations

Polonceau Truss

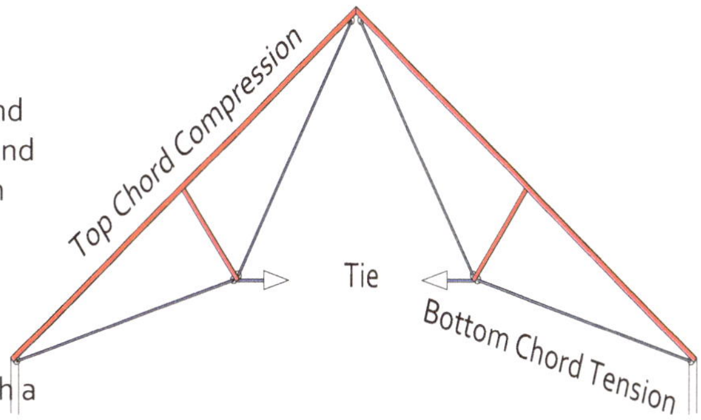

Another historic truss that originated as a compound of timber and iron rods (timber as principal rafters and rods as ties), which was highly successful in the 19th century.

Many examples survive to this day. In essence, the Polonceau truss, shown in **figure 2.6.18** comprises two kingpost trusses leaning against each other with a central tie preventing spreading at the eaves.

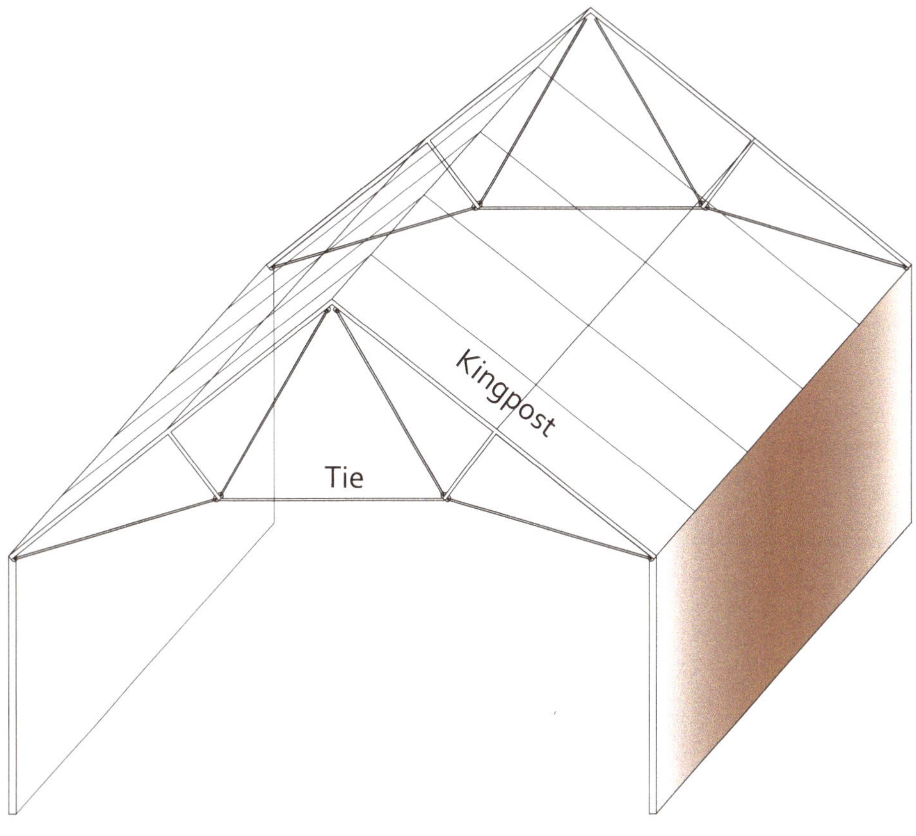

Figure 2.6.18 Polonceau Truss

Vierendeel Girders

This style of truss was named after the Belgian civil engineer Arthur Vierendeel, who developed the design in 1896.

In principle, a Vierendeel girder works in a way similar to a truss, except that the sloping ties or struts are omitted. The compound beam action is achieved by means of frame action, which transfers internal bending moments between the elements.

Figure 2.6.19 illustrates in diagrammatic form a vierendeel.

A number of vierendeel bridges were built in the early 20th century, but many were destroyed in the two world wars. They tended to be more expensive than other options.

Vierendeel girders have a special place in building construction. Taking advantage of the floor-to-floor height in a building a very deep and efficient, which avoids any cross bracing obscuring openings for windows or doors.

An example is shown in **figure 2.6.20**, which was used to carry a floor and roof over an 8m external opening.

Solid cladding panels were inserted between the posts.

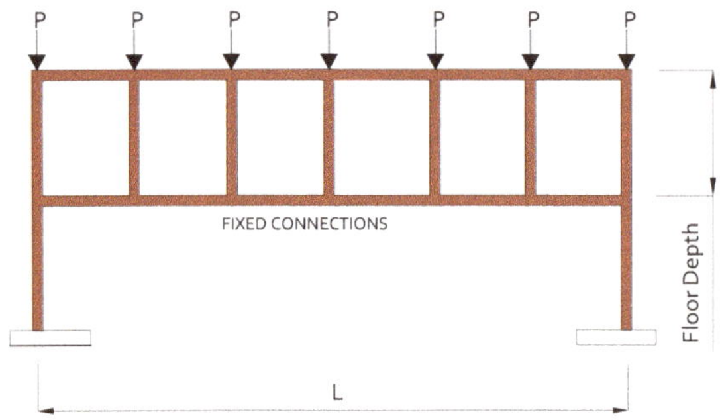

Figure 2.6.19 Virendeel Frame

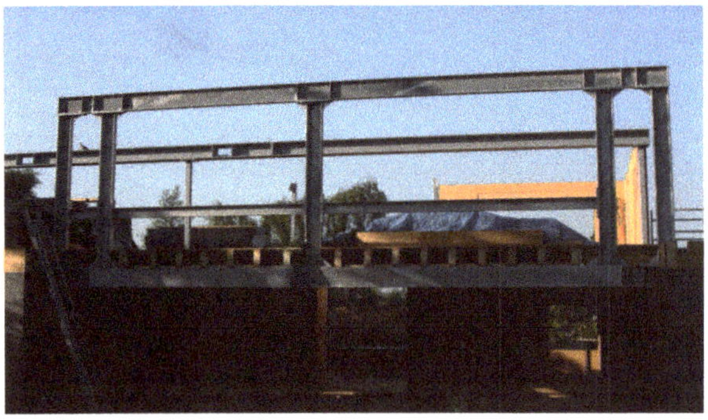

Figure 2.6.20 Virendeel Frame Example

Analysis of a Vierendeel Girder

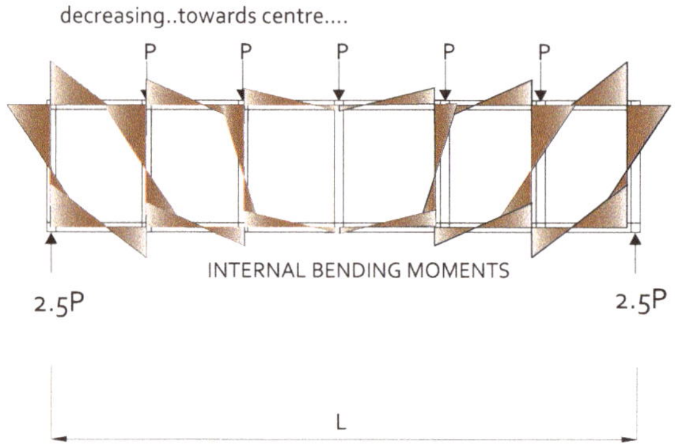

Figure 2.6.21 Vierendeel Internal Moments

The principal means of absorbing the applied loads is by means of internal bending at the fixed joints internally. **Figure 2.6.21** indicates the internal bending moments within the frame. The bending moments are very similar to the wind frame in the previous chapter (**figure 2.5.11**) but turned on its side.

It is of note (like the wind frame) that the loading will accumulate towards the support point. The internal bending moments increase from the centre of the span to a maximum at the support.

This is a slightly confusing concept: the point of maximum internal bending moments occurs at the point of overall maximum shear. In a simply supported beam with a UDL the point of maximum shear is the point where there is minimum external bending moment.

Assuming that all of the internal elements are the same section size, then the worst case design is the end strut. Referring to **figure 2.6.22** the end moment within the strut may be estimated, in this configuration, as 25% of the resolved moment (end shear x bay width).

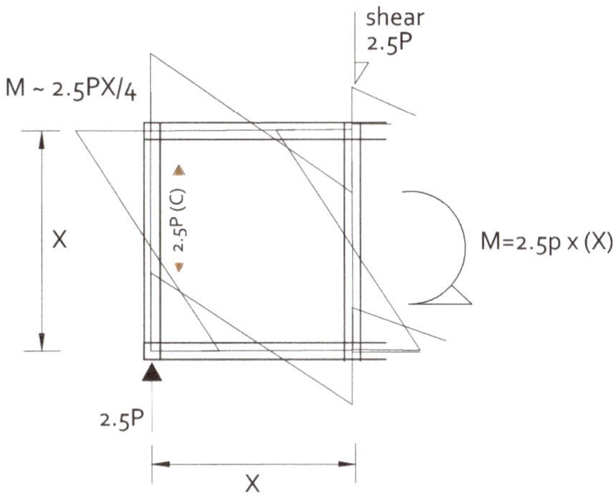

Figure 2.6.22 Vierendeel Support Moments

Note that this element also has a compression loading in it, which equates to the support reaction. When designing this element, the compression and bending will need to be checked together.

Like a truss, the centre of the span is a location where the top and bottom chord experience maximum tension and compression due to bending, as illustrated in **Figure 2.6.23**.

In order to estimate the stiffness a similar method to that described in **figure 2.6.10**, in the truss section, may be used.

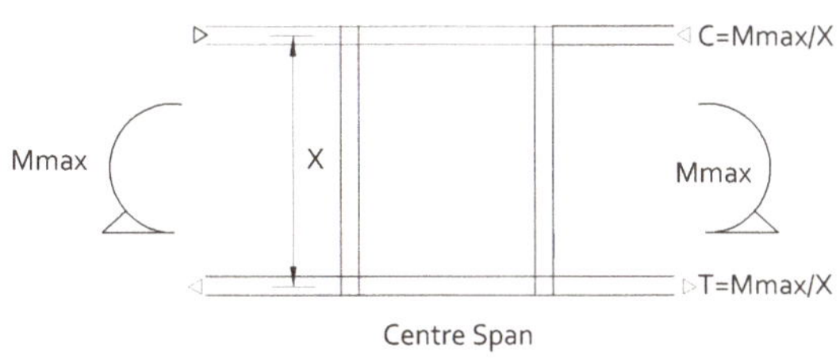

Figure 2.6.23 Vierendeel Midspan Chord Forces

Stiffness Estimate

A cautionary note is that the assumed stiffness is likely to be less than the 'quasi' beam depending on the internal element stiffness – a 2D computer model is the most expedient method of assessing this effect.

Assessing the deflection can be readily undertaken following the initial sizing of elements by resolution of estimated moments adjacent to the support. It may demonstrate that the use of the same section size for elements in the centre of the span is needed to limit deflections, despite their lower fixing moments.

Sir John Deanes College

Deep Trusses and Vierendeel Girder

Trusses and vierendeel girders may be combined to span large spaces – particularly when utilising the deep floor-to-floor space in a building. **Figures 2.6.24** and **2.6.25** illustrate how a dramatic overhang was formed in a college building. The architect designed a dramatic entrance. In order to support the overhang, two trusses were designed in the façade, at right angles to the corner, with backspans onto internal columns.

The truss to the front edge needed intermediate support. An internal truss would have impeded the internal library space. The solution was a vierendeel, in which the vertical posts were integrated into the library space without the need for internal struts. It was supported by an edge column and a backspan.

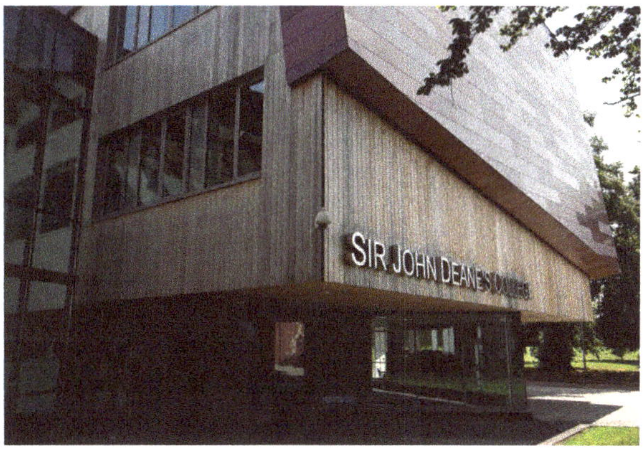

Figure 2.6.24 SJDC Dramatic Entrance Overhang

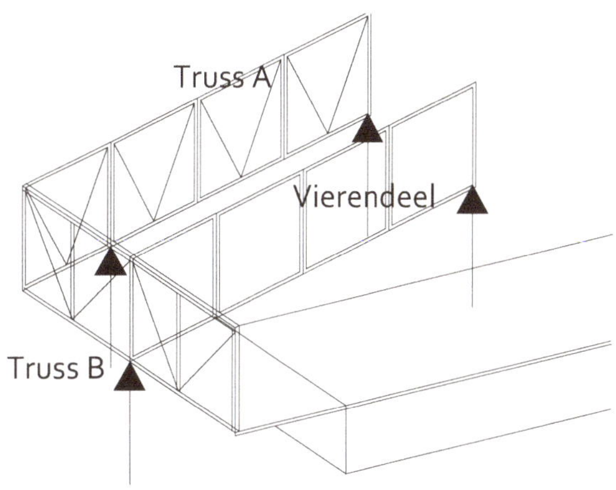

Figure 2.6.25 Combined Trusses and Vierendeel

Chapter 2.6

Quick Note Summary

1. 1. Trusses act as beams – the top and bottom chords, like the flanges of an I-section beam and the internal struts and ties as the web.

2. A quick way to estimate the element sizes in a truss is to resolve the external (applied bending moment) in the centre and to divide this by the distance between the chords in the centre. The strut/tie size is estimated by resolving the support reaction with the first internal strut.

3. A useful method to estimate internal forces is to form a virtual cut through the truss and resolve the resultant quasi-external balancing forces.

4. The stiffness of a truss may be estimated by calculating an idealised stiffness using the areas of the top and bottom chords (ΣAy2). Because many trusses are 'deep' in relation to their span, a more accurate assessment will entail some form of shear deformation checks.

5. Kingpost, Queenpost, and trussed timber joists are hybrid structures: a compound beam and truss.

6. Roof trusses have been used in various forms: the Kingspan, Queenspan, and Polonceau are common historic examples.

7. Vierendeel trusses work by means of internal bending moments at the fixed joints between elements.Struts and ties are omitted.

8. An estimate of the element sizes of Vierendeel girders may be resolved by using the support reactions with the first internal framework.

9. Estimated deflections of a Vierendeel girder may be found using a similar method to 4 above. Beware of the flexibility of internal elements causing less than expected stiffness.

Chapter 2.7

2.7 Tension and Suspension Structures

This chapter investigates the properties and use of tensile fabric and suspension structures in buildings with initial references to bridges, which are more commonly associated with this form of structure. The starting principle of catenary action and resolution of a suspension bridge cable forces is followed by the example of a real stadium roof at Chelsea Football Club. A cable stay bridging structure at Lancaster University is reviewed together with a suspended walkway in Manchester, UK. The chapter concludes with an examination of simplified methods of analysis of tensile roof structures.

A contemporary form of a suspension rope bridge in **Figure 2.7.1** has wire cables slung across, from which hangs a suspended deck.

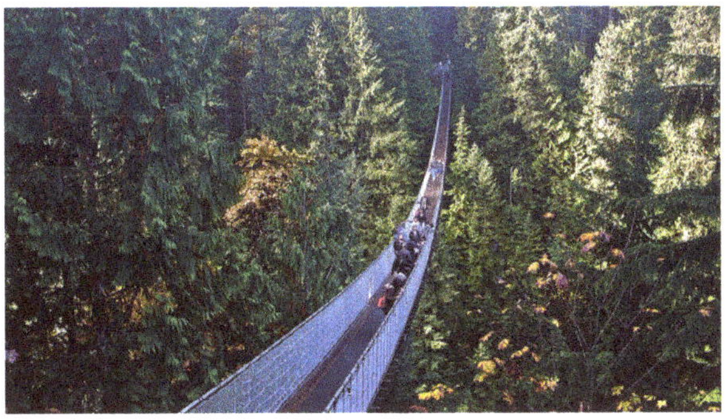

Figure 2.7.1 Capillano Suspension (Rope) Bridge

The suspension bridge is a sophisticated modern derivative of the rope bridge.

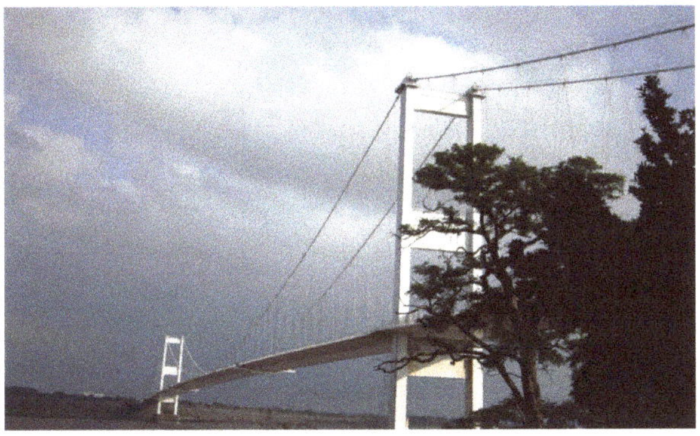

Figure 2.7.2 Suspension Bridge

These bridges are capable of spanning great distances (up to 1-2km) and carrying road traffic. The ropes have evolved into thick, high-tensile cables wound with high-tensile steel, and the decks have post-tensioned concrete or steel hollow sections. The decks are often designed aerodynamically to avoid wind turbulence causing resonance, and the towers cast in slip-formed concrete or built from steel.

Figure 2.7.2 shows the Severn Bridge in the UK. Perhaps the most renowned of this genre is the Golden Gate Bridge in San Francisco, USA.

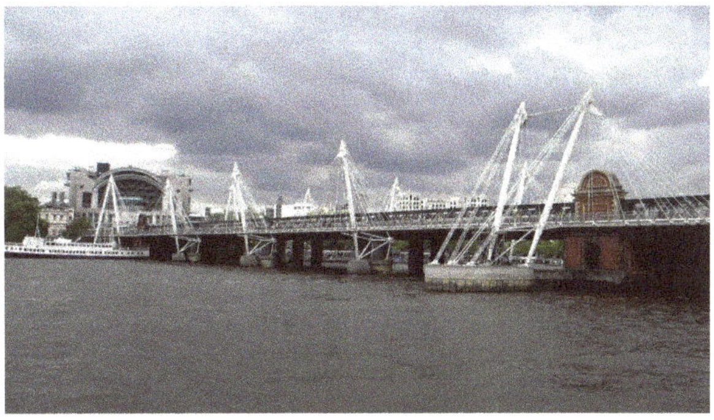

Figure 2.7.3 Cable Stayed Bridge

Vying with the suspension bridge as a long-span structure is the cable stay bridge. See **Figure 2.7.3**.

The deck is carried by high tensile cables or bars, which fan out from the main support towers – multiple suspension supports rather than a single, catenary shape.

What is the Benefit of Tension Structures?

Put simply, tension is the most direct and efficient way to carry loads prior to the ready availability of ductile steel historic structures comprised mainly of masonry and timber. Timber was in use for its availability and flexural properties; although it was also used for tension as a tie – **see figure 2.6.16** (kingpost).

With the advent of steel rods and cables and formerly wrought iron chains, tension structures were possible. Without having to worry about buckling failure, only the direct load path needs be considered, with slender sections being possible. Slenderness and elegance is a descriptions often associated with tensile structures.

The Catenary

The catenary is a very useful phenomenon: it is the shape that a cable or chain takes up when hanging between two fixed points. A more mathematical description is the *at-rest* position of a chain, which takes up the least potential energy.

The shape transcribed is a hyperbolic shape, similar to a parabola, the formula that took some time for mathematicians to solve.

Figure 2.7.4 shows three stands with linked cables, the sort that are used in queuing barriers in banks. The shape of the cable is intuitive: the shorter the distance between the posts, the deeper the curve.
The weights at the base of the posts help to stop horizontal thrust from the chains toppling the posts.

A cable or chain may defined as an infinitely flexible structural element that carries loads by means of internal tension only.

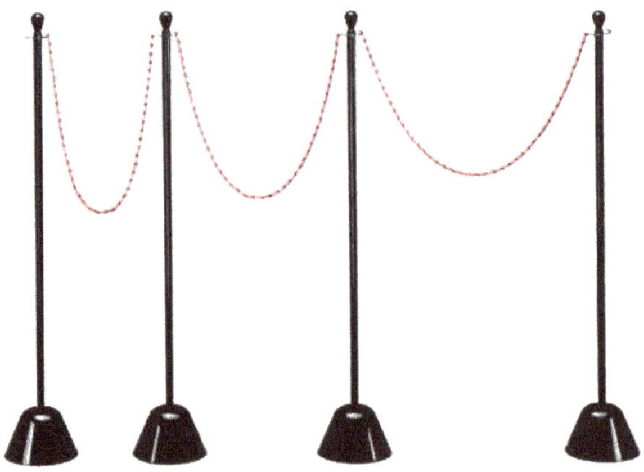

Figure 2.7.4 Catenaries

In **figure 2.7.5** the deformed shape is determined, not with any reference to the deflection of the simply supported beam (wl4/384EI), but with a mathematical description related to the relationship h α cosh x (h is proportional to cosh x), where cosh is the hyperbolic cosine of 'x.'

It is also of note that the supports cannot be fixed: rotation of the cable may always occur at its support.

Robert Hooke helped Sir Christopher Wren to find a suitable form for the dome of St Paul's in London as an inverted form of the catenary cable: that is a dome.

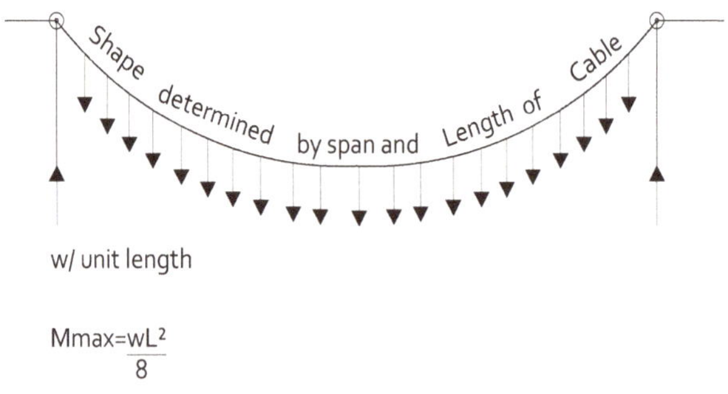

Figure 2.7.5 Catenary Cable

The total bending moment on the simply supported system is the same as for a simply supported beam, but with no flexural capacity to carry a load, the chain or cable assumes a shape that is similar to a parabolic bending moment diagram in a simply supported beam.

From Catenary to Suspension Structure

Figures 2.7.2 and **2.7.8** illustrate how dramatic structures are possible when making use of catenary action.

When significant loads are carried, then it is not possible to carry a loaded catenary or suspension cable with simple posts and counterweighted feet, as in **figure 2.7.4**.

The load of the catenary collects at the supports, as will be demonstrated in the next example, whereby the engineer must find a way to anchor the cable, whether with foundations or with a structure designed for the purpose, such as the roof end trusses in Chelsea Stadium in **figure 2.7.8**.

There is scope for ingenuity here, but the engineer should have a clear appreciation of how to anchor the end thrusts.

The next examples will show how to calculate in approximate form the forces in a cable and the end reactions.

Suspension Bridge Forces

Attention to the classic suspension bridge is a good introduction to long-span lightweight roofs in other applications.

A suspension bridge comprises a relatively stiff deck hanging from a catenary cable. The cable has negligible rigidity. It does not need to be designed with respect to bending moments internally. The deck, which needs to be rigid enough to carry heavy loads, has a bending stiffness and strength that may be designed to span between suspension points rather than the full span L. See **Figure 2.7.6**.

'To Engineer is Human' (Petroski 1992, reference 8, chapter 13) makes particular reference to suspension bridge failures and problems. The cause of failures frequently relates to the deck design which should be designed to mitigate the effects of side winds or sway induced by traffic.

Remedies involve making the decks stiff, introducing dynamic damping, or shaping them aerodynamically.

A simple resolution of forces is possible from a suspended cable, which is of great use for estimating cable sizing and support reactions. Returning to the simply supported condition, the total bending moment carried between supports over span **L** is **$WL^2/8$**.

The horizontal reaction is related to the rise (h) of the cable, the span (**L**), and the load carried (**w** per unit length).

From the horizontal reaction, the anchoring cable tension (**T**) may be found. See **Figure 2.7.7**. In other words, the necessary anchoring forces by which the foundations must hold the cable in place may be estimated very simply. The designer may decide on the design of the foundation needed to anchor the cable.

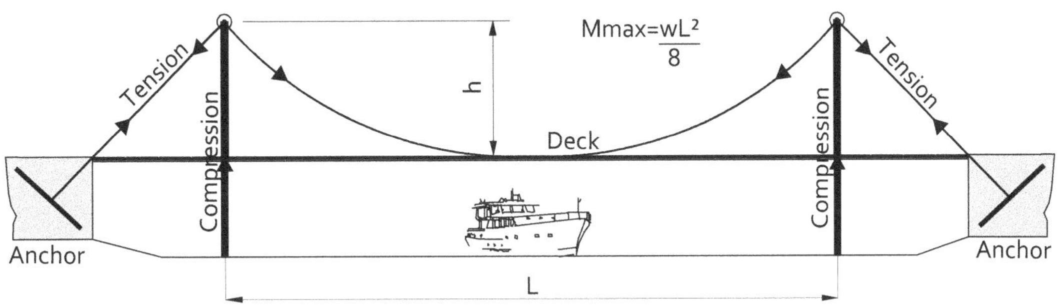

Figure 2.7.6 Suspension Bridge Forces

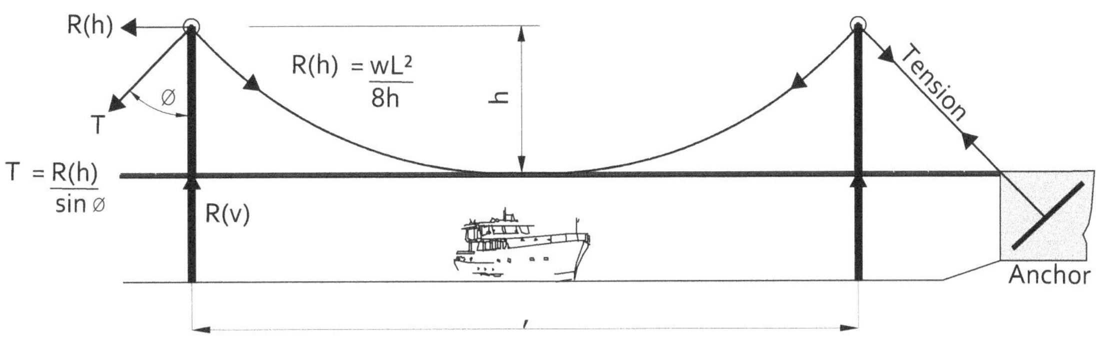

Figure 2.7.7 Suspension Bridge Horizontal Reaction

Chelsea Football Club Catenary Roof

The inspiration of Stephen Morley was to adopt a catenary solution to support the North and South Stand roofs of Chelsea stadium. **Figure 2.7.8** shows an overhead view of the South Stand roof in which the overhead roof beams hang from a catenary (in this case, steel tubes with flange joints). The catenary is anchored at the ends by trusses, which restrain horizontal end reactions so that they are self-contained. The vertical forces are carried to the ground by columns to the perimeter of the stand.

One of the corner trusses supports a 15-tonne television suspended beneath.

Many stadia roofs are supported by cantilever structures. This ingenious roof made use of the fact that Stamford Bridge Ground has limited space in which to form a back span for a cantilever.

The catenary roof is particularly efficient and economical on two counts:

1. Referring to **figure 2.3.3** in the Cantilever chapter, the bending moments and deflections for cantilever beams are respectively 4 and 10 times greater for a given structure than for the equivalent simply supported span.

2. Support is provided by a catenary in tension (pinned releases at the tube joints ensured the catenary was designed without internal bending in the elements). Tension cables are more efficient than structures designed for bending or compression - which would need to be designed to be robust enough not to buckle under load.

Stadium roofs are an area of design where the structural engineer may design elegant structures that may become icons as familiar to the public as the Golden Gate Bridge or Forth Bridge. The exaggerated arched form of Wembley Stadium has become just one such icon.

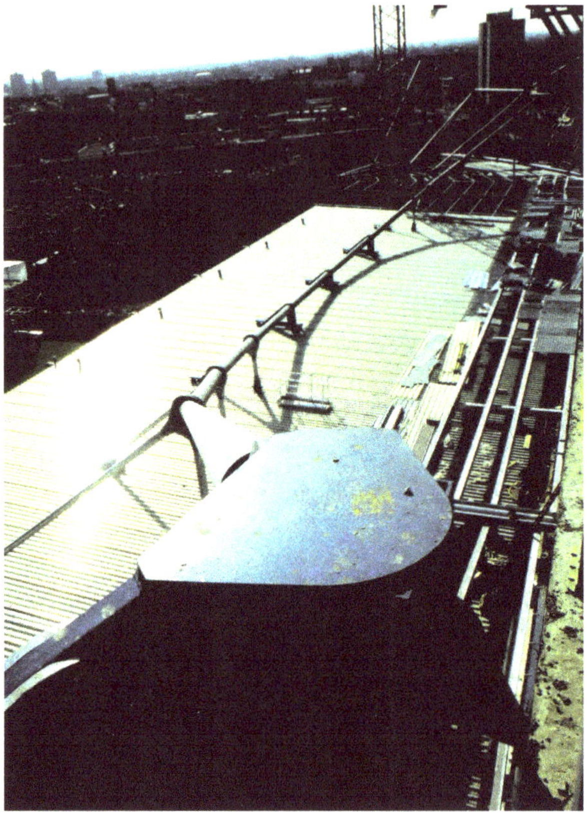

Figure 2.7.8 Chelsea South Stand Roof

An interesting aspect of the roof in **figure 2.7.8** is that, in the plane of the roof, there are steel tubes that follow the line (like the shadow reflection) of the catenary cable. This tube is fixed between the roof beams. The reason for this element is that the roof catenary leans at an angle to the vertical, thereby exerting horizontal loads, which are in turn balanced by the roof plan arch.

This form of structure is rather like the Brunel railway bridge, which spans over the river Tamar in Devon, UK. Tubular wrought iron arches span over the top of the roof deck, the end reactions being balanced by an inverted arch underneath. Both are original means of balancing the catenary loading.

The next chapter contains a worked example of an arched stadium roof (**figure 2.8.24**), the Golpharm stadium in Huddersfield, which is an inverted form of a suspension cable.

Cable Stayed Bridging Structure

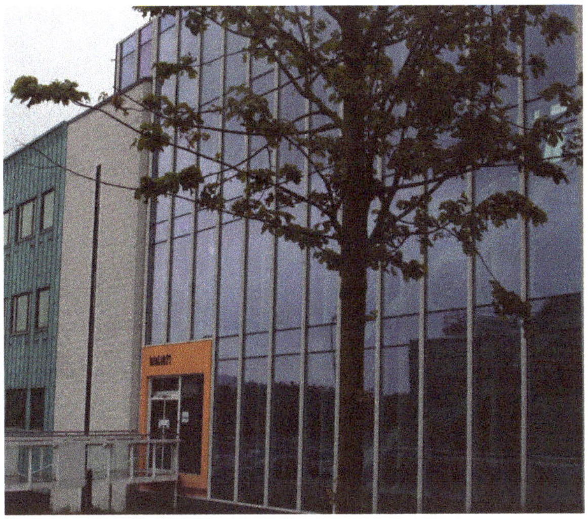

Figure 2.7.9 Cable Stay Entrance Infolab

At Infolab, Lancaster University, in **Figure 2.7.9**, the entrance foyer to the building was designed to bridge over an 18m pedestrian underpass. Suspension rods are expressed behind the glass façade. The lobby is the equivalent of three stories in depth. Tensile forces in the bars enabled them to be thinned down without having to consider buckling.

A simplified conceptual force diagram is shown in **Figure 2.7.10**. One of the ties was omitted at the entrance door to allow space for the door; an architectural compromise, but the principle of the force diagram is unaffected.

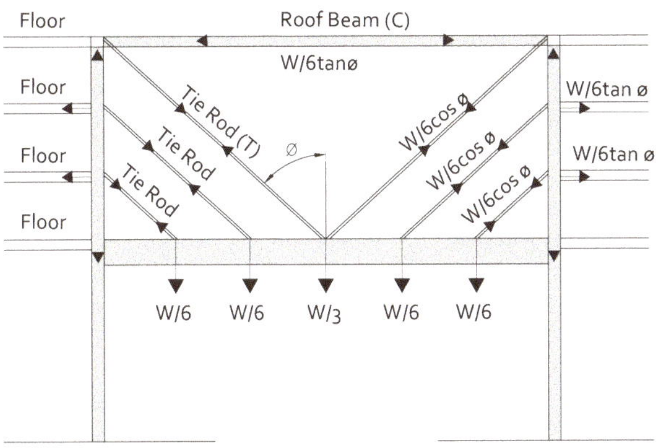

Figure 2.7.10 Infolab Bridge Simple Force Diagram

The bridge structure supports the floors across the width of the lobby utilising the full depth of the façade. It was checked with a self-contained 3-dimensional model. This two-storey structure is carried by vertical columns (omitted for clarity) to the lower deck.

The loads distributed across the deck are supported.

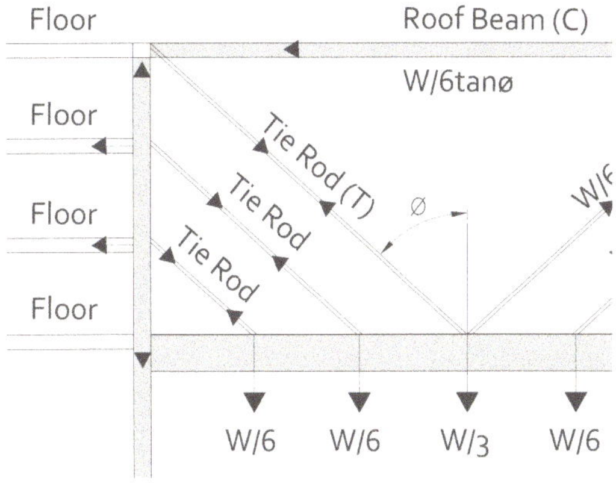

Figure 2.7.10 (part view)

by the raking (*sloping*) hangers. As in a truss, the internal forces of the top hangers are reacted by a roof beam with a compression load, and the two lower hangers by the floors, which tie back the whole structure.

Deflections had to be checked carefully; they were limited by the amount of movement that the window frames could accommodate.

Figure 2.7.11 shows the base detail for the tie rods and the columns at the façade. The tapered fork connections for the columns and the ties express well the forces that are transmitted through the connections.

Fork ends are often available ready fabricated, but architects often like to design purpose-made end fittings. This is one part of the design process where engineers may actively play a part in assisting with the selection of aesthetically pleasing details.

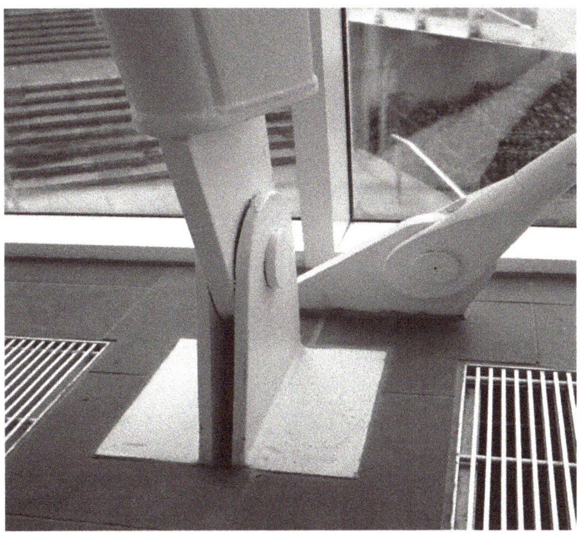

Figure 2.7.11 Hanger Detail

Suspended Walkway

In **chapter 3.1** the collapse of suspended walkways at the Hyatt Regency disaster is discussed in some detail. A solution for six stories of suspended walkways for an apartment building atrium in Manchester was to suspend one edge of the walkway from beams from roof beams. See **Figure 2.7.12**.

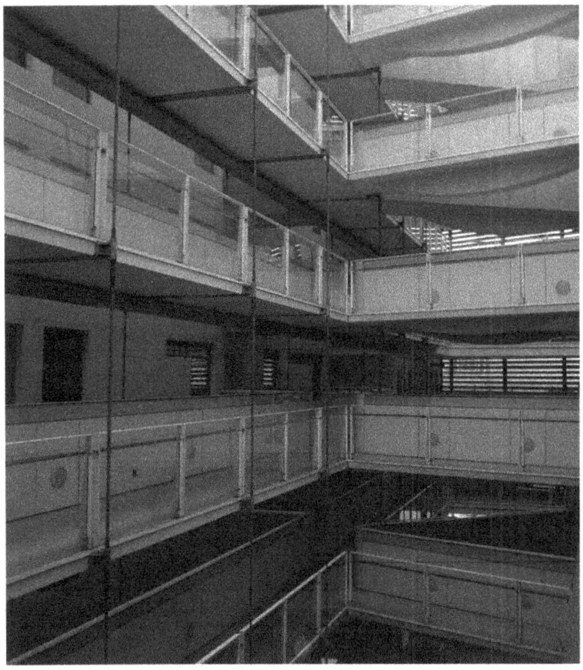

Figure 2.7.12 The Locks Internal Walkway

It is arguable whether the detail shown in **Figure 2.7.13** was any more desirable aesthetically than the Hyatt Regency walkways, but a similar principle occurs. All six stories are suspended from the roof.

The topmost walkway hanger bar carries all of the stories above.

Hanging structures are a way of creating slender supports that do not have to extend to the ground floor. Visually, they help to give an impression of open-plan spaces and enable, in this building, a ground-floor atrium space that is unimpeded by internal columns below the walkways.

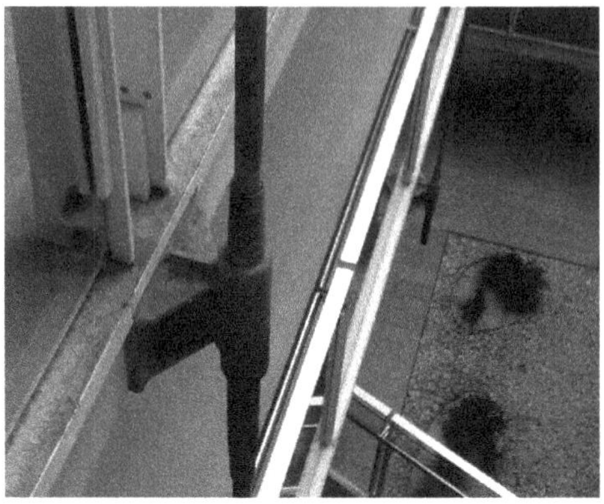

Figure 2.7.13 Hanger Detail

At the concept stage, a quick assessment of the loading is easily conducted; see the calculation below.

Calculation of Approximate Tie Force:

Area of Floor carried by bar per floor:

1/2 floor width x distance between bars 1.5m^2

Weight of floor carried (Self Weight and Live load is 5kN/m^2):

Area carried x Unit Load = 5x1.5 = **7.5kN**

Total load carried by top bar over 6 stories:

6 x 7.5 kN = **45kN**

Select Diameter of Bar

Tension Capacity of bar = 275 N/mm^2)
.....ignoring thread area:

Factor up load (ultimate limit state):
Tension approx., T= 45x1.5 = 67.5kN

Bar area = $\pi D^2/4$ (Circular Section)

Maximum tensile stress in bar = T/A
= $4T/\pi D^2$ = 275N/mm^2

Diameter of bar must be greater than
$\sqrt{(67.5\times10^3\times4/\pi/275)}$ = **18mm**

The actual bars measure approximately 25mm, so allowing for the loss of area when cutting the thread, they are sized appropriately.

Figure 2.7.14 shows a conceptual loading diagram. Six stories of walkways hang from a roof beam.

The hanger load is transmitted from floor to floor through the edge hanger.

The connection from which each walkway hangs from the bar must transmit the loading from just one walkway at each point.

There is no accumulation of force in the hanger bracket itself. See **Figure 3.1.2** as a comparison Hyatt Regency Hotel failure.

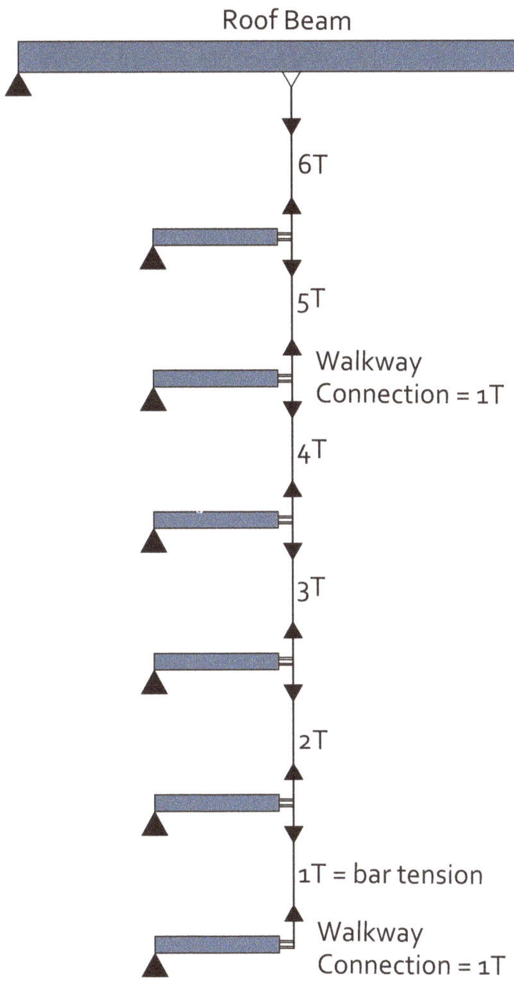

Figure 2.7.14 Walkway Loading The Locks

Aside

Construction Sequence

Manchester Locks - Construction Sequence Tension structures are elegant, lightweight, and may be used to great effect. However, they introduce special problems for the contractors building them.

IIn the Locks building, the roof beams had to be built first (in order to hang the walkways). This caused two particular problems:

1. ***During construction, as the building rises, the usual (bottom-up) form of construction would allow building the walkways at the same time as each floor, thus allowing the walkways to be used for access for trades fitting out (decorating and building the internal finishes).***

2. ***Having built the roof beams, there is an immediate issue: the support beams themselves may impede lifting in of the walkway decks: this creates a headache for the contractor: can the walkway decks be built in segments, or can the tie rods be put together from roof level and hung down as 6 segments connected from which the walkways may be supported?***

3. ***What about health and safety issues, lifting in large assemblies in an area where many trades are working together in congested circumstances?***

4. ***What about temporary propping requirements? Does the contractor have to bear the cost of installing temporary structures to support the edges of walkways whilst the tie rods are being fixed.***

Tension Fabric Structures

Frei Otto was one of the pioneers of this genre of structures, which are a magnificent addition to the family of engineering structures.

These are essentially tented structures, a sophisticated form whereby the shape is computer-defined (an idealised shape with no internal bending like a chain), and the material is tied to strengthening cables or too stiff edges (beams or walls).

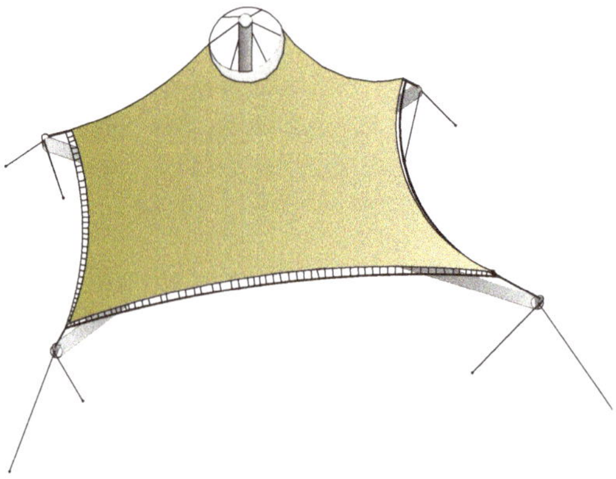

Figure 2.7.15 Tension Roof Structure

Another form of structure is realised by stretching the fabric over arched beams or tubes. The fabric bends tightly over the beam, forming a pleasing curvature.

This section will look at a symmetrical example where a mast holds up a funnel-shaped fabric canopy that is spread onto radiating arms (or steel struts/booms).

Structures that are akin to a modern dome tent – whereby fabric is stretched over curved beams - are omitted in this section.

Figure 2.7.15 illustrates the roof to be examined, and **Figure 2.7.16** shows a PVC tension roof at Goodwood Racecourse, a series of conical mast bays. This was designed on the principles to be discussed in this section.

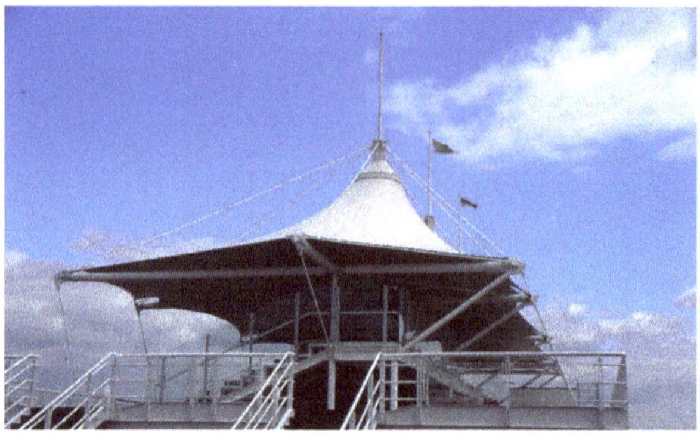

Figure 2.7.16 PVC Tension Roof Goodwood Racecourse

Analysis of the roof in **Figures 2.7.17** and **18** would generally be undertaken by a specialist using a *non -linear* analysis program which will also *form-find* the shape of the roof in a process known as dynamic relaxation. This is an iterative trial-and-error form of analysis.

Engineers with no experience in these structures tend to hand them to specialist designers. But there is no reason why they cannot be scheme-designed by a competent engineer with a knowledge of the proposed geometry.

Simple Method of Scheme Design

A design procedure might be as follows:

1. Sketch a framework and desired roof profile roughly to scale. Decide on the support structure and holding down points. There may be preliminary drawings from the architect to guide the engineer at this point.
2. Decide on the fabric. From this, a typical pre-stress fabric tension may be assumed. Typical fabrics and pre-stress values are tabulated in the next few pages.
3. Find the perimeter cable tensile forces based on their geometry.
4. Resolve at the junction of perimeter cables and holding down ties for external reactions.
5. Balance external reactions to find the supporting mast or opposite compression element loads.

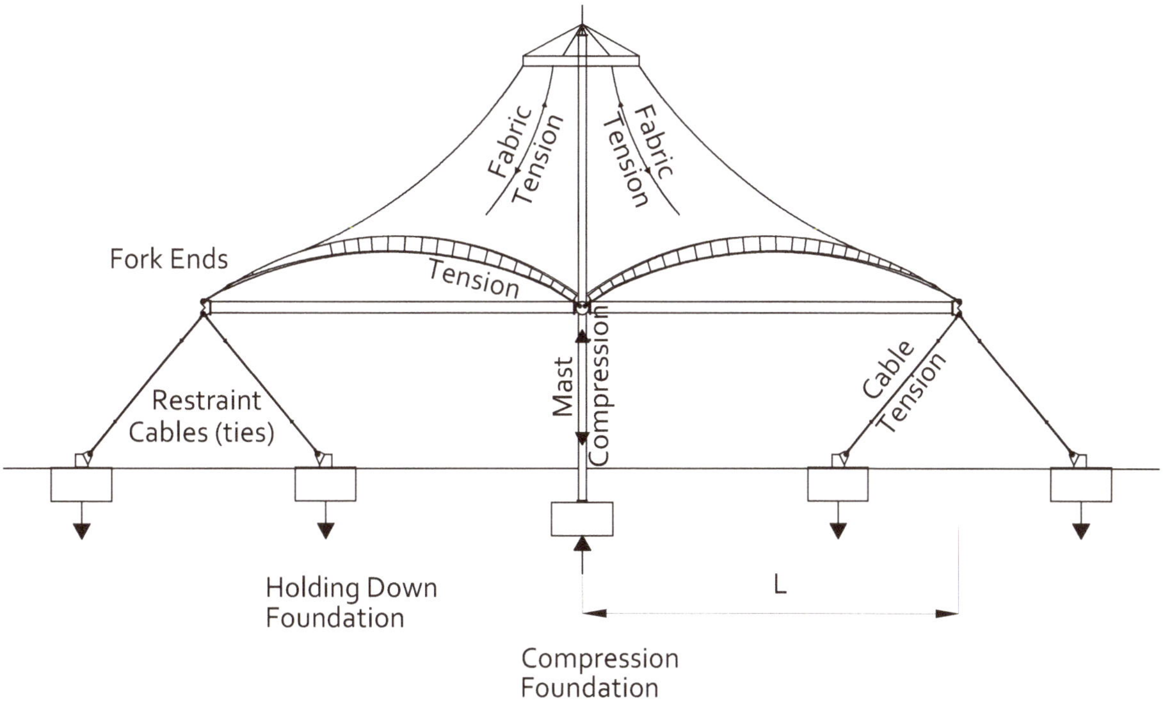

Figure 2.7.17 Force Diagram

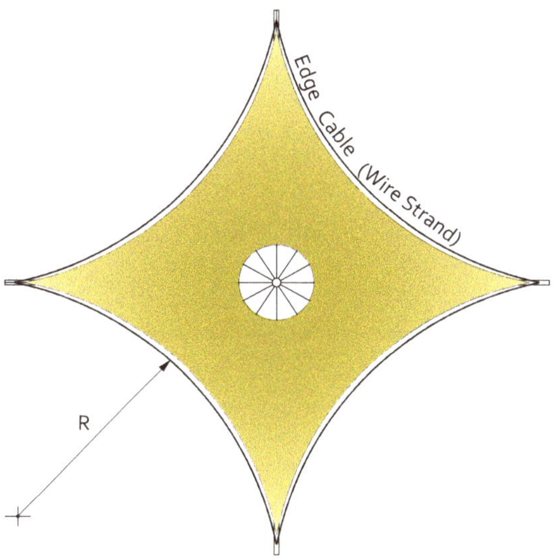

Figure 2.7.18 Plan View

Conceptual Design Method

Based on the previous procedure simple approximate methods combined with a linear model of the cables and supporting structure can enable the concept designer to size elements and cables and to try out different shapes and configurations in a short time.

Figures 2.7.17 and **2.7.18** show a symmetrical structure comprising masts and booms (yachting nomenclature is commonly used). A tension fabric (PVC) is suspended from a ring beam at the head of the mast and pulled into a curved form out to four corner points at the tips of the booms.

The booms are prevented from being pulled upwards by a pair of cable ties at each end. The cable ties in pairs will also act to prevent any side winds from forcing the structure over.

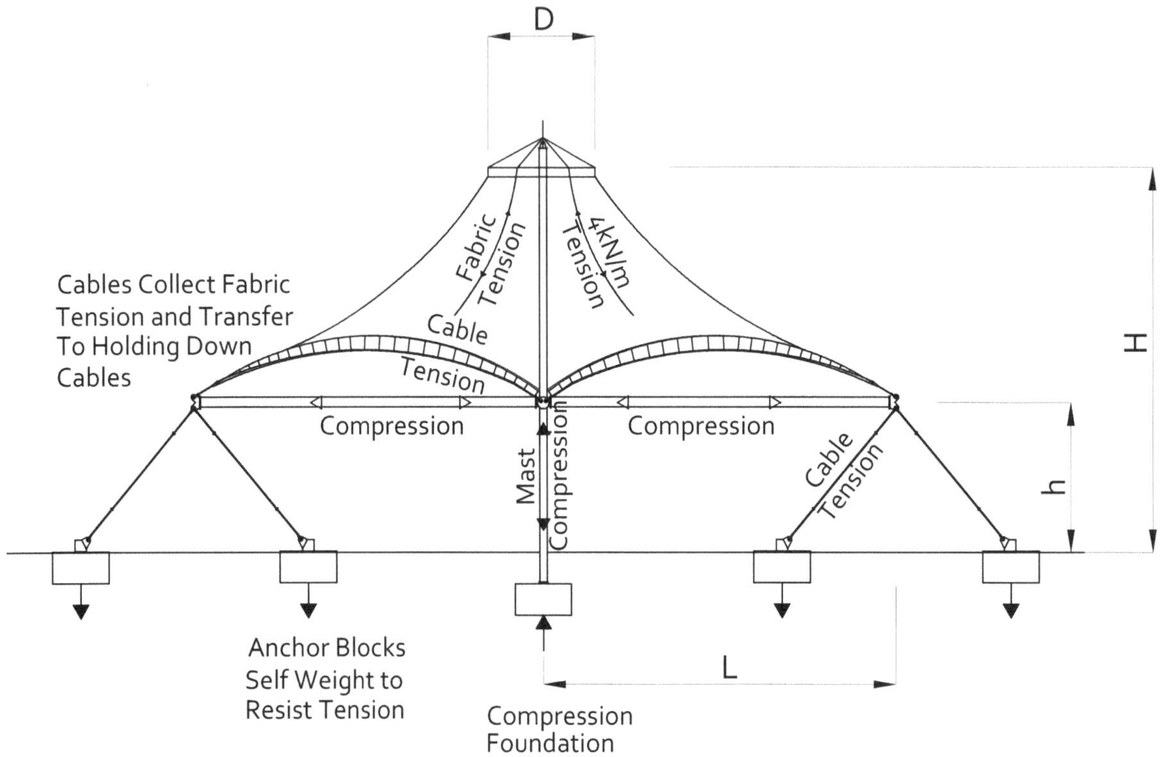

Figure 2.7.19 At Rest Force Diagram

Fabric Types

The most commonly used fabrics are:

- PVC – *good overall properties and hard-wearing, but not such longevity as PTFE.*
- Glass Fibre PTFE coated – *strong and durable but susceptible to tearing.*
- ETFE (film) – *translucent, Inflated cells – framed, not tension fabric structures hitherto described.*

Tension Fabrics Working Pre-stress (tension)

PVC - 1-4kN/m (note more or less the same in warp and weft)

PTFE - 1-2kN/m - light fabric
4-6kN/m - medium fabric
6-8kN/m - heavy fabric

Non Linearity

The beauty of the approximate design of tension structures is that the stretched internal load is used; non-linearity is not a concern. A constant fabric stress is assumed, and the loading is resolved locally thereof.

Aside

Stresses are described as 'warp' (in the main direction of fibres – downwards direction in the roof - figure 2.7.17) and 'weft'. This is analogous to the directional span of plates (see chapter 2.10). Warp and weft stresses are not necessarily equal. This can be investigated in a detailed non-linear analysis. But for the purposes of the approximate cable force calculation, which follows a warp line load may be assumed.

Example Structure Calculation

Viewing **figure 2.7.20** and **2.7.21**, relating to boom end horizontal forces, there are two stages:

1. Assess Internal Forces and size elements – from 'at rest' condition with fabric in tension (mast and post).
2. Assess External Forces (Wind, Lateral impact etc.)

Main Parameters (ref. Figure 2.7.19)

Total height of structure = 6m (H)

Height to the horizontal booms = 3m (h)

Length of booms (horizontal struts) = 4m (L)

Fabric (PVC) tension = 3kN/m *fully stressed*

Diameter of upper ring beam = 2m (D)

Radius measured in plane, edge cables = 10m

Calculate Edge Cable Tension (ref. Figure 2.7.20)

For a cable of radius **R** (assuming perfect circular curve), with radial tension applied, the tension is given by.

Tension *(warp and weft)* in fabric = **T**:

Tension in cable = T.R
= 3kN/m x 10
= Tc = **30kN**

Calculate Boom Compression Force

Figure 2.7.20 shows on plan the cables junction with the boom (the cable linking boom and main edge cables is ignored for simplicity).

Resolving on plan, (Tc-cable tension)

Boom force = 2Tc.cos45
= 2x30xcos45
= **42kN**

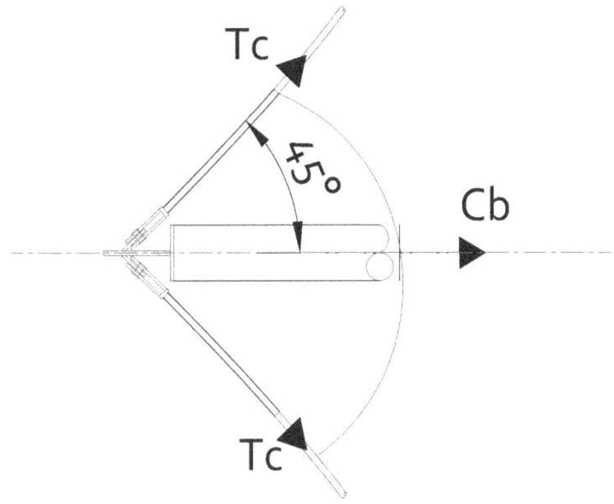

Figure 2.7.20 Boom End Horizontal Forces

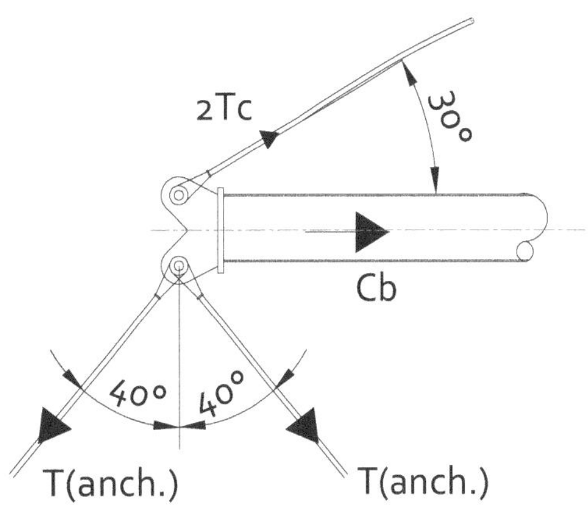

Figure 2.7.21 Boom End

Calculate Tie Down Forces

The main cables which tie down the end of the booms to the ground is found from vertical resolution of forces:

Resolving vertically (see **figure 2.7.21**)...

2Tc sin 30 = 2T(anchor) cos 40

Therefore....T(anchor)
= Tc. sin 30/cos 40
= 30 x 0.5/0.77 = **20kN**

Uplift on Tie Down Blocks (concrete footings under anchors)

Uplift (vertical)
= T(anch.) cos 40
= 20 x0.77 = **15kN**

.....under each anchor point

Size concrete foundation

Work out weight of concrete to hold down the anchor (with 50% added safety factor):

Weight required = 1.5x15 = **22.5kN**
(2.25 tonnes)

Density of concrete = 24kN/m^3

Assuming a 1mx1m square concrete base.....

Depth of concrete block needed
= 23/(24 x 1m x 1m) = **1m deep**

Calculate Compression in Post/Mast

This assumes equilibrium of the post compression and uplift forces on the blocks:

Total of tie down forces (calculated previously)

4 tie down points at 15kN = 4x15 = **60kN**
Unfactored mast load = 60/1.5 = **40kN**

This is the compression in the post.

Calculate Fabric Stresses at the Cone

Σ(sum) line loading in upper ring = **πDT**, where T is the tension line load and πD is the circumferential length.

In this example the fabric is nearly vertical at the top of the mast. If it inclines at an angle to the vertical the post load is the resolution of forces; as in a truss node.

Say, angle of slope to horizontal is 60°, then the force in upper ring is πDT sin60° (πDT cosØ where Ø is angle to vertical). The fabric load is found using the column compression.

Diameter, D = 2m

Assuming 60°, line load in fabric at the ring
= 40kN/π/ 2m/sin60° = **7.35 kN/m**

This is fairly high and may require widening the ring or doubling up the fabric (PVC has a strength of around 60kN/m-110kN/m excluding any safety factor, so it may not be likely). Internal cables may also be used to reinforce the fabric.

Specialist software undertakes form finding of the fabric to optimise the shape, allowing the designer to alter the edge parameters. The 'found form' will tend towards a shape with more even stresses. The purpose of the exercise was to illustrate how the defining parameters may be investigated with approximate calculations.

Wind Loading

Wind loading is an external force. In other words, when the wind blows onto the structure, there is a net sideways and upwards or downwards pressure acting on the whole structure. This is resisted by the structure with reactions in the mast and ties down to ground.

Figure 2.7.23 shows wind loading acting on the roof. Given the shape, there is a tendency for overall suction. This will tend to lift up the fabric, thereby relieving the internal tension. When assessing the wind loading, the internal tensions may be ignored initially by resolving for the wind acting. A rough general design wind pressure is 1-1.5kPa (kN/m^2).

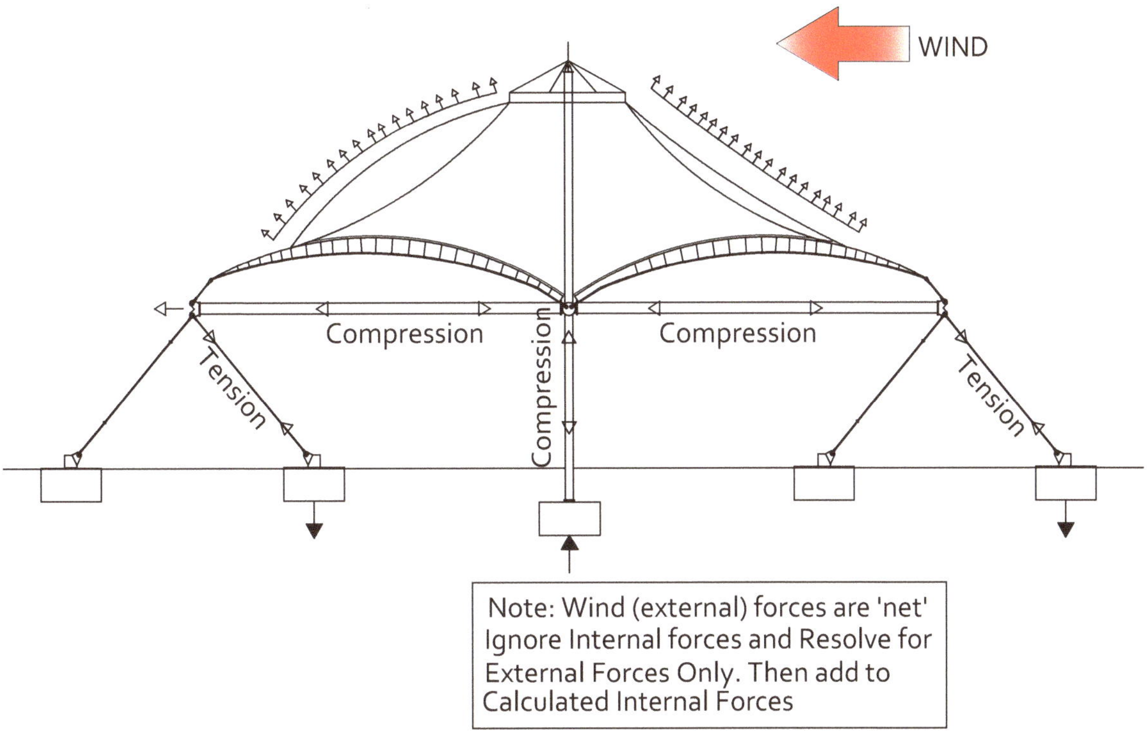

Figure 2.7.22 Wind Force Diagram

Approximate method of calculating wind forces:

1. Calculate the face area of the roof, which is facing the direction of the wind.
2. Multiply by the design wind load.
3. Assume this force (WxA) is applied to the relevant booms facing the wind.
4. Resolve horizontally into the two cables shown in **figure 2.7.22** (two cables share the load in this case).
5. 5. Check whether external cables are needed to assist in reacting wind loads

(See **figure 2.7.22**).

Load Combination

Comparing *at rest* and *wind loading* cases the main effect of a combination is to increase the boom loading and the anchor tension.

The total anchor and boom forces are now obtained by summing the tension and compression, respectively, from the two cases.

Aside

Boom and mast sizing: both elements are compression members. Consideration of size needs to take into account the curvature under load of slender members. There are various modern formulae (such as the Perry Robertson), but the simplest formula for sizing, useful for approximate calculations was developed by Euler in 1744.

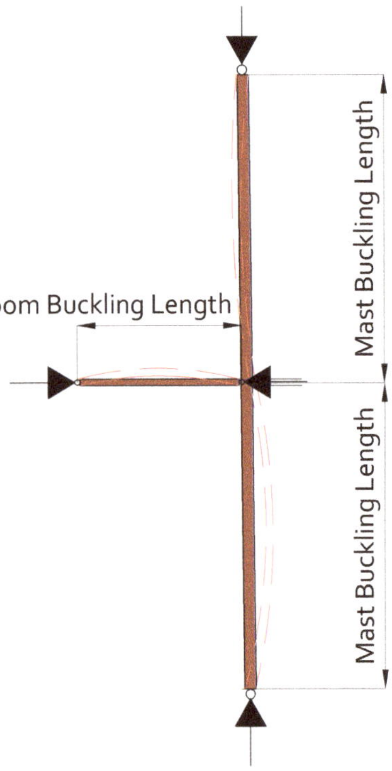

Figure 2.7.23 Euler Buckling Diagram

To calculate the Euler critical buckling load:

Locate points of restraint (for the mast half way up, for the boom the full length) This gives the buckling, or effective length (Le).

Select 'K' multiplying factor for strut. K=1.0 for pinned at both ends and K=2.0 for one end pinned and the other free (1.5 for a cantilever - one end fixed one end free).

Calculate Pcr or Euler critical compression force:

$P_{cr} = \pi^2 EI / Le^2$

E and I are the modulus and moment of inertia (minimum) respectively.

When using Euler a second check should be made, which is to calculate the actual critical compressive load based in yield compressive stress:

Critical compression load, **$P_c = \sigma_y \times A$**

Where σ_y = yield compressive stress for the material, **A** = area of section

For a section which is not slender this is generally less than the Euler buckling load, **Pcr**.

Edge Cable Detail

There are two common methods to link the edge cable to the fabric. An aluminium clamping plate is shown below in **2.7.24**. Bolts are tightened onto the plate over the fabric edge and cable to hold the two sides in a friction grip.

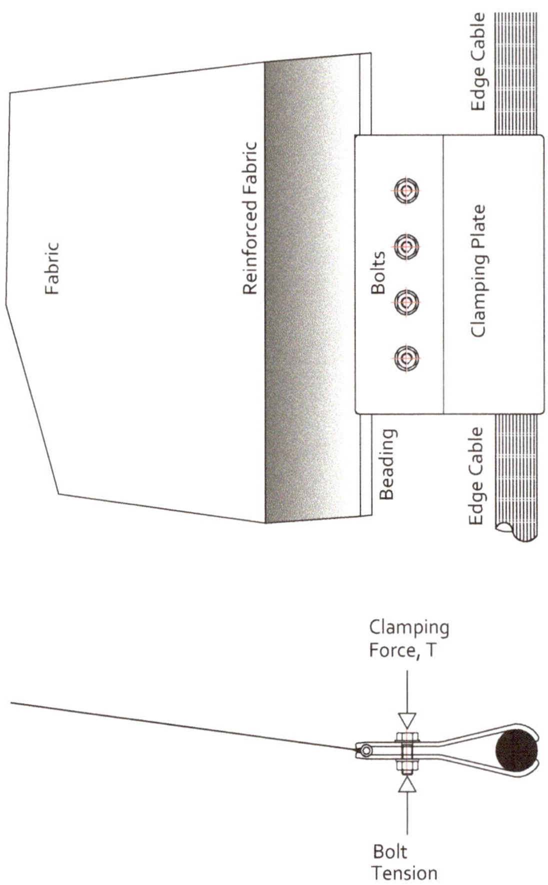

Figure 2.7.24 Edge Cable Clamping Detail

Calculation of Bolt Tightening Torque

In the following "aside" example, a useful approximate calculation that relates the tightening torque in a bolt to its pre-load is a useful means of enabling the engineer to specify bolt tightening.

A tool called a torque wrench may be used by an erector to check bolt-tightening torques on site to ensure that the cable clamps are sufficiently fixed. It is of note that hand-tightened bolts are limited to an applied force of up to 200N (20kg). If a spanner is assumed to have a length of 250mm, for instance (0.25m) then the applied maximum torque is in the region of:

Hand Applied torque limit = 200x0.25 = 50Nm

Aside

How to Calculate Tightening Torque

It is often necessary with friction grip type details to specify a tightening torque for a bolted fixing. A special device called a torque wrench is used by the fitter to ensure that the specified torque is reached.

Overdoing the torque can shear the bolt.

Many drilled anchor fixings specify torques. There is insufficient use of torque wrenches in the industry to check torques, which give the most surety that a fixing has been correctly installed.

A method of estimating the torque required for a specific bolt tension is shown next. This uses a simple empirical formula.

Note that the constant is based on SI units. This needs to be modified for empirical units.

Calculation of torque...

Hand tightening can achieve torques in the region of 50Nm..

If a wrench is 0.2m long and is tightening a bolt, a person with reasonable force may achieve a force approaching 100kgf (100N)...

This is equivalent to a torque:
0.2mx100 = **20Nm**

80Nm would be on the high end of achievable torques using hand held wrenches.

Relationship of torque to bolt pre-load...

A simple empirical equation in SI units is:

Simplified Pre-load Equation

Pre-load (T) = 5 τ / D

Where τ is the torque in Nm, D is the bolt diameter in mm and T is the pre-load in kN.

A 12mm diameter bolt with a torque of 50Nm has a pre-load of 21kN

This is greater than the tensile capacity of a 4.6 grade black bolt, so a stronger bolt would be needed to achieve this torque.

The coefficient on friction, μ, varies for different types of surface (0.1-0.2 steel-steel up to 0.40 against a rough concrete face).

Assuming a μ value of 0.2 with a safety factor of 3.0 against slip, the above bolt tension will resist a lateral force of:

Lateral resistance = Tension (bolt) x μ /3.0
= 21x0.2/3 = **1.4kN**

Chapter 2.7

Quick Note Summary

1. Suspension bridges and Cable stay structures are an efficient way to span large distances.
2. Catenaries are a *pure* elliptical or hyperbolic curve.
3. There is no bending in a cable or fixity possible at catenary support.
4. The horizontal reaction for a suspension cable at the support point approximates to wL2/8h – where h is the dip distance of the cable.
5. The tension in the tie-back cable of a suspension bridge is possible by resolving the forces (as described in the truss section) – the nearest slope angle for the suspension cable is the tangent at the support.
6. Wide-span roof structures are possible with the use of tension rods and tall columns.
7. Cable-stayed type supports with raking (or sloping rods or cables) may be adopted in building structures.
8. Ties for multi-storey suspended structures are sized for cumulative loading.
9. There are two main types of tension roofs: mast type and frame type (fabric stretched over curved arches or domes).
10. Tension roof structure support frame may be sized by assuming a tension in the fabric and resolving forces (1-4kN/m PVC and 4-6kN/m PTFE).
11. For example, edge cable forces are estimated by the equation *Tension = Line load x radius of the cable*.
12. Wind loading effects on a tension structure may be found by analysing the wind and ignoring the internal tension initially. The wind and internal forces may be combined to create a *worst-case* envelope.

Chapter 2.8

2.8 Arches, Vaults and Domes

Compression structures are as ubiquitous as traditional bridges, cathedrals, and viaducts. They have been adopted in modern forms using steel, concrete, and timber. Traditional forms of vault and arch are reviewed, leading from a simple bridge with a single arch to multiple arches and to a simplified analysis of a cathedral-buttressed section. Finally, the reader is invited to try out the methods of simple analysis to size a steel arch structure (based on a real stadium roof). By way of this example, step by step conceptual design shortcuts are followed to illustrate the way that a practising engineer might conceive a design without the use of a computer.

Arches, vaults, and domes are traditional structures that have been used for millennia to construct bridges, cathedrals, and viaducts.

A particular characteristic is that they may be constructed readily from masonry and stonework, which may be sourced locally and require no special fabrication aside from quarrying and cutting to shape by stone masons. As long as they conform to particular geometric proportions, many have remained stable for hundreds of years, even in earthquake zones.

From the 19th century, more modern forms of structure include reinforced and pre-stressed concrete, steel, and wrought iron. Some shell structures which were designed by engineers such as Eladio Dieste have employed brickwork and concrete in mathematically defined shapes as thin shells which are highly efficient with very economical use of material.

Within this chapter, design assessments are highly simplified. 'The Structural Skeleton' (Heyman 1995, reference 5.) is recommended for a wider understanding of the principles of structural action.

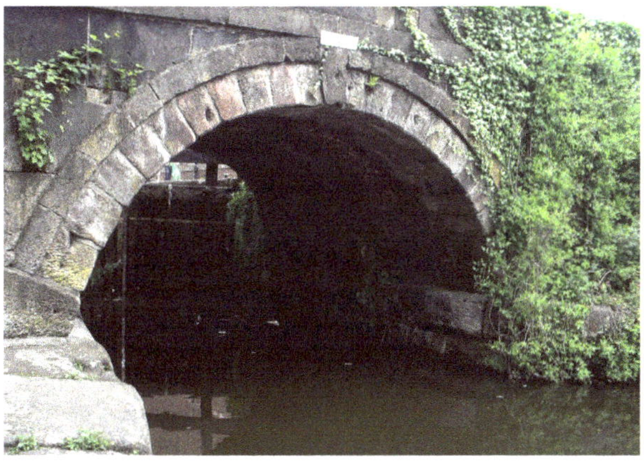

Figure 2.8.1 Traditional Canal Arch Manchester

The Arch

In the previous chapter, the catenary was described in which the pure tensile shape of a structure may be found by suspending a chain between two points.

Turning the shape thus inscribed upside down will define the mirror image: the pure arch form, which is catenary (elliptical) in shape.

Arches, in practice, do not have to conform to this shape. Many are semi-circular, or for the traditional Muslim architecture or medieval churches, they follow a pointed form. Within solid deep masonry structures, they may follow their own thrust line, or in thinner vaulted structures, they can form hinges within the structure, allowing the non-parabolic shape to work effectively.

A reason for the endurance of masonry and stone arches is the fact that the internal compressive stresses are usually low in comparison with the strength of the material. The supporting abutments may settle or rotate, and the arches or vaults or domes may form cracks internally – throwing the internal forces into a different load path, finding a new state of equilibrium.

The assembly of a traditional arch construction is outlined in **figure 2.8.2**. **Figures 2.8.1** and **2.8.3** show this form of construction.

The lower-most stones which form the arch are called voussoirs. They are cut accurately to interlock, meeting in the centre of the span at the keystone. They may not necessarily carry the principle thrust of the arch in compression, but they define the shape and, by their interlock, hold in place the supported masonry or stone to prevent slippage.

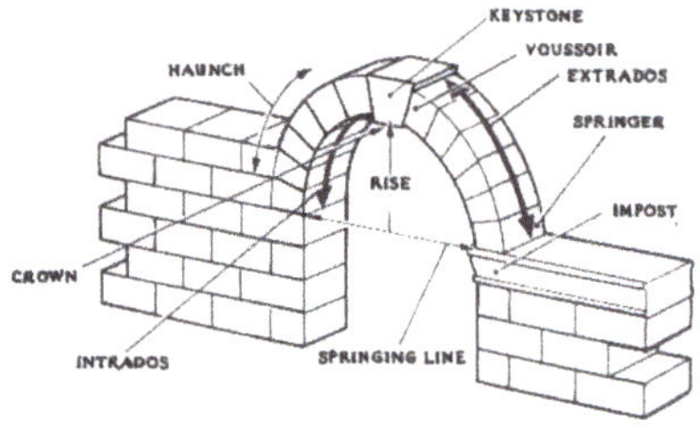

Figure 2.8.2 Arch Components

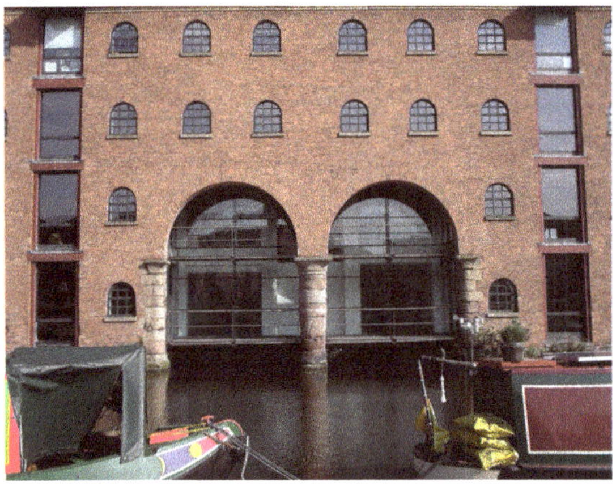

Figure 2.8.3 Merchants Warehouse Manchester

Analysis of An Arch

The following **figures 2.8.4** and **2.8.5** are an apt illustration of the thrust line (the path of the compressive force) in an arch-approaching collapse.

There are two sets of imposed forces in the bridge:

1. Self-weight. This is more or less uniform compression acting across the arch. Like the catenary cable, the self-weight causes a line of thrust, which is essentially compression acting in a curved shape that approximates to the shape of the voussoirs. The compression force presses the blocks against each other.
2. A load is applied, as shown in **figure 2.8.4**. - say a heavy vehicle. As the vehicle moves across the bridge, the vertical loading causes a thrust line which compresses the voussoirs. The greater the force or weight of the vehicle, the more the thrust line will move downwards at the middle span and upwards at the third span points

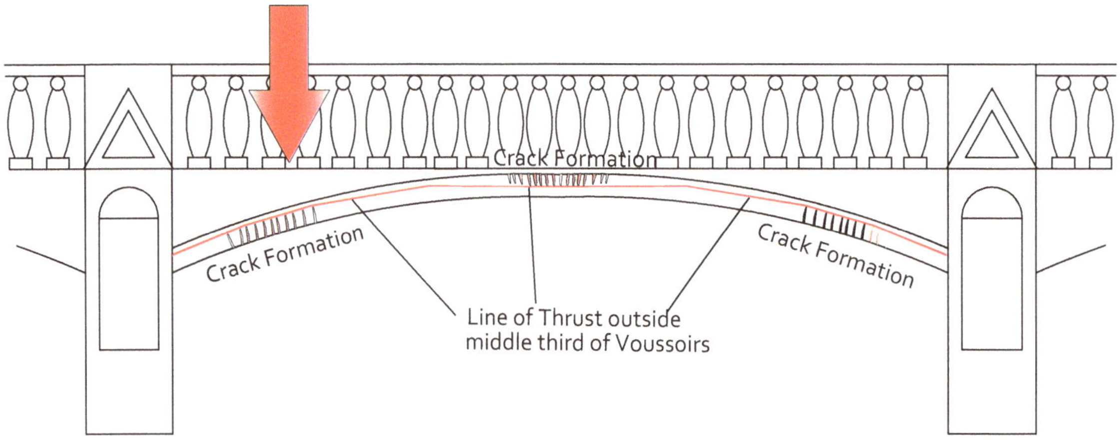

Figure 2.8.4 Three Pinned Arch Crack Formation

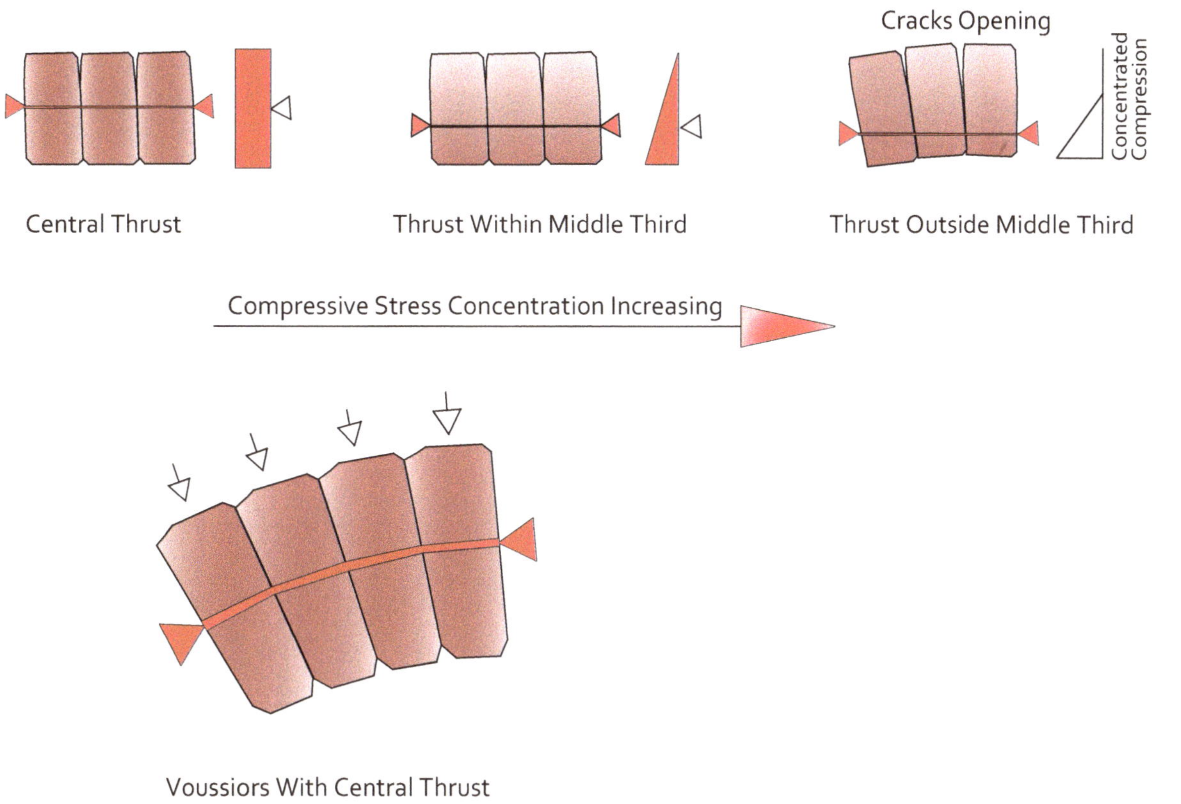

Figure 2.8.5 Thrust Line across Voussoirs

Progressive Stresses for Increasing Loading

Imagining an increasing load being applied to the arch above, the thrust line will move as indicated in **figure 2.8.5** – from left to right.

A similar shape of the thrust line occurs under foundation bases exerting pressure on the ground.

The stacked blocks represent the voussoirs as the load increases the thrust line slides from the centre to the middle third of the blocks.

The compressive stress block diagram is shown below: it changes from a rectangular (constant) shape which then becomes triangular (compressive stresses peak at the inner edge).

When the load is in the middle third the blocks have a compressive stress across their width.

At the collapse point the thrust line moves outside the middle third. The stones pull apart at the outer edge. Since it is masonry no tension may form. At a critical point the stones may crush causing the bridge to collapse.

Window Lintels

This form of structure was traditionally used from window heads; a repetitive architectural feature to the façade of buildings.

Two variations are shown in the extract from 'The Building Educator' (Greenhalgh, reference 4). See **figure 2.8.6**. These arches were constructed with temporary supports (or *centring*) in timber, which were *struck* (or removed) on completion.

Today they are relatively rare, with proprietary galvanised or stainless steel frames being designed and used for permanent support with permanent centring designed to carry the load.

As a reflection on modern building methods, the craftsmanship and knowledge of construction that go together to build flat arches and other traditional types of structures, such as the cantilevered stair, have largely disappeared. A challenge, perhaps for the modern engineer.

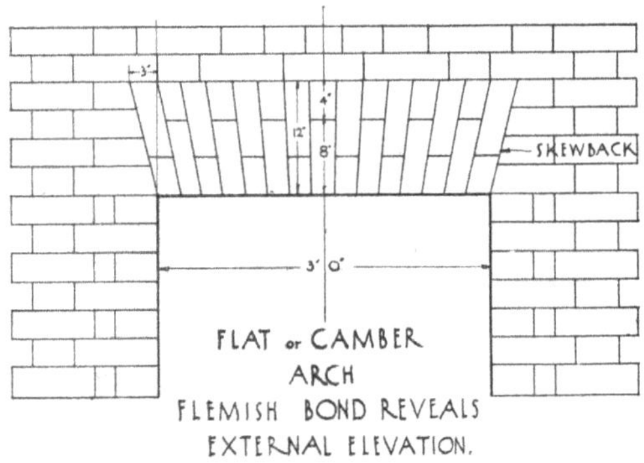

Fig. 57

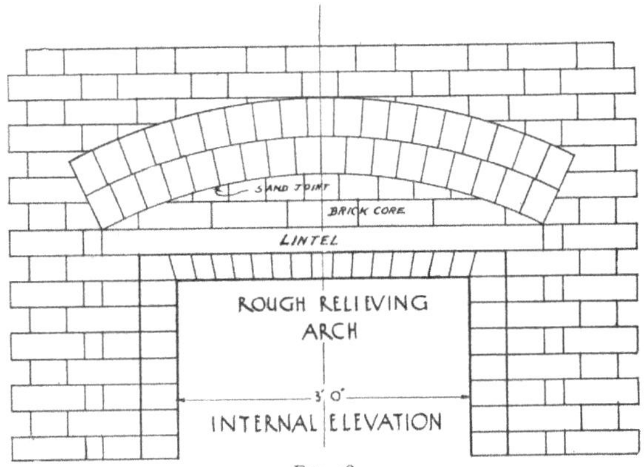

Fig. 58

Aside

Middle Third Rule

The middle third rule is of great value for compression problems, in particular for compression within a concrete or masonry construction and for foundation pads compressing on the ground (see chapter 2.13; page 150-151). If in doubt with a compression problem, it is always worthwhile to keep the centre of loading within the middle third of a compression block.

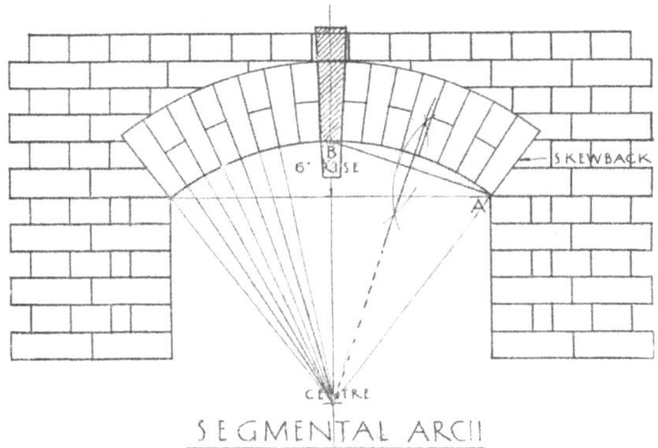

Fig. 54

Figure 2.8.6 Window Arch Lintels

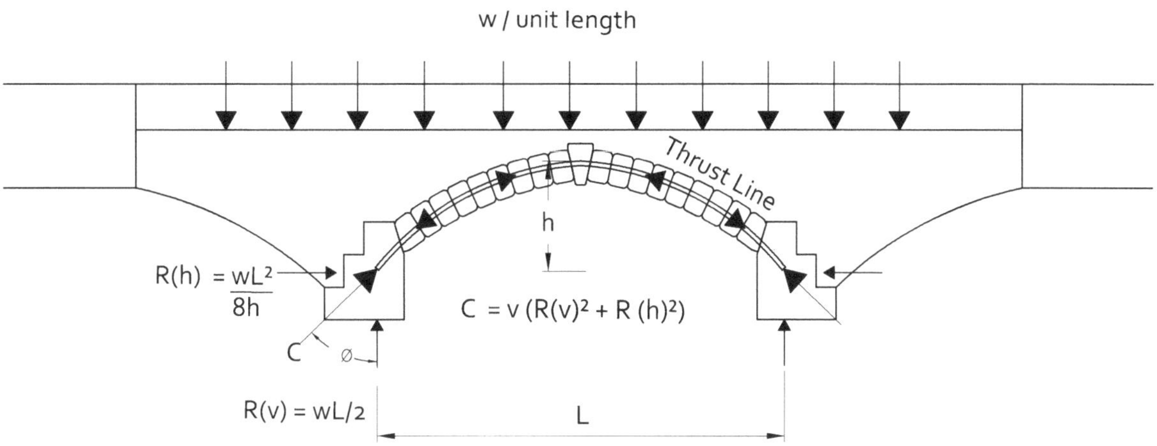

Figure 2.8.7 Arch With Uniform Load

Arch with Uniform Load

Under uniform loading, if it is assumed that the thrust line assumes a perfect parabolic form, the resolution of forces in an arch with uniform loading is shown in **figure 2.8.7**. Like the suspension cable an approximation of the thrust in the arch may be found by finding the horizontal and vertical reactions and resolving these into the arch compression thrust force.

Approximation of Loads

The following calculation is illustrative of elementary calculations, which make it possible to estimate forces in a bridge in the unloaded state.

Assuming, say, half the depth of the voussoirs above the stress in the stone blocks may then be found.

Finding Voussoir Compressive Stress

Arch width = 3m
Voussior block depth = 450mm
Weight of Bridge = 50kN/m

- equivalent to about 1m height of stonework over the voussoirs.

Voussior Compressive Stress Continued...

Span (L) = 30m
Rise (h) = 10m

Horizontal reaction = $50\times30^2/8/10$
= 563kN R(h)

Vertical reaction = 50x30/2
= 750kN R(v)

Compression Thrust in Arch
= $\sqrt{(563^2 + 750^2)}$
= **937kN**

Compression stress in stone assuming half depth,

Compression stress
= $937\times10^3/3000/(450/2)$
= **1.39 N/mm² (or MPa)**

If the bridge is sandstone with a quoted crushing strength between 6.3 MPa and 214 MPa (N/mm2), the actual stress is about 17% of the lower bound sandstone strength. In general, masonry arch bridges are lightly stressed in comparison with their strength.

Assessing Bridge Capacity Using Hinges

Unlike politicians, engineers are mostly employed to provide a definitive (yes or no) answer. Imagine that a local highway authority wishes to assess the loading capacity of the bridge example, in **figure 2.8.7**, what is the maximum weight of a single vehicle that may be carried by the bridge?

Suppose a dedicated team of bridge engineers is available with the latest finite element modelling techniques. But how can the engineer make an initial assessment to find the capacity of the bridge?

A simple method is analogous to the truss analysis in **figure 2.6.8.** The arch is assumed to reach failure point where the *thrust line*, falling outside the middle third of the voussoirs, causes crushing of the stone.

It cracks open in a pattern similar to **figure 2.8.4**: the centre span with cracking forming on the upper surface (intrados) of the *voussiors* (internally) and location adjacent to the support on the outer surface (extrados).

Forces are resolved, as for the truss, in a virtual cut section (or free-body diagram). Internal thrust forces are represented as balancing external loads.

Force Resolution at Failure Point

At failure, hinge points form (a hinge at the supports and two in the middle), noting that four hinges need to be formed for instability. A three-hinge bridge is stable.

For simplicity the engineer might sketch the bridge to scale and select three equal segments of equal weight with hinges at third span points. See **Figure 2.8.8**.

As in **figure 2.6.8** (trusses), a free-body diagram is selected with balancing external forces on the left-hand side. See **figure 2.8.9**.

At the point of collapse that, the horizontal reaction R(ha) is assumed to be located as indicated in **figure 2.8.9**. This reduces the number of equations to resolve to a single equation.

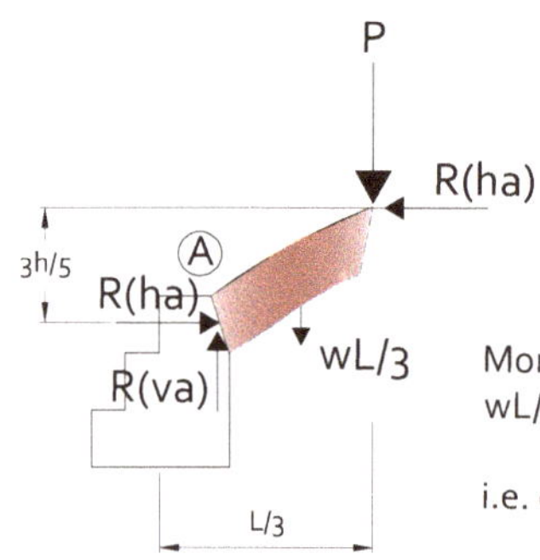

Moments about A:
$wL/3(L/6) + P.L/3 = R(ha).3h/5$

i.e. can solve for R(ha)

Figure 2.8.9 Free Body Diagram at LHS Support

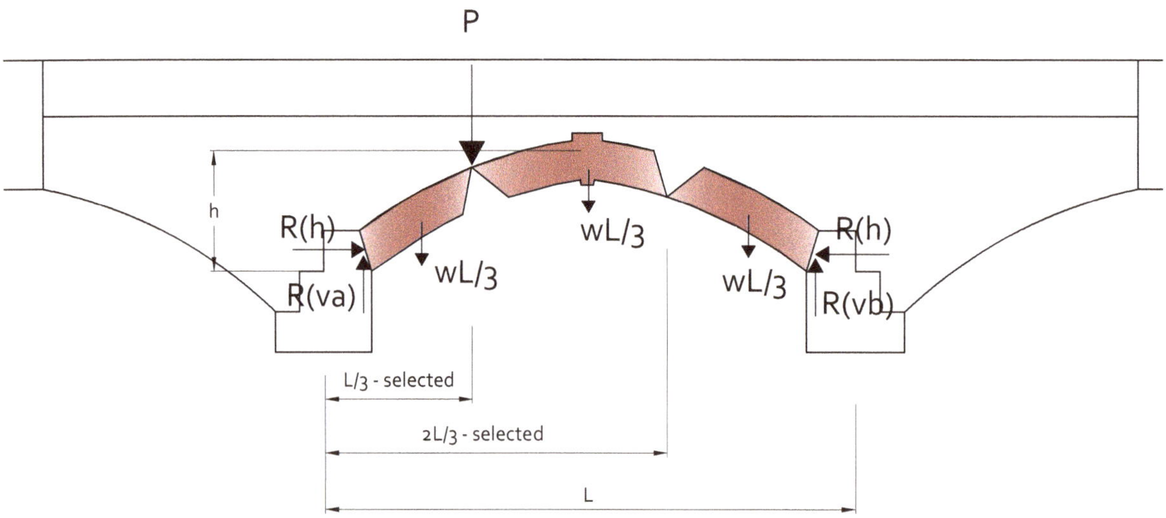

Figure 2.8.8 Arch: Three Selected Hinges

Assumptions

When undertaking approximate calculations, it is a cardinal rule to lay out, or at least be clear about, the assumptions made:

1. Arch thrust = **R**(ha), horizontal reaction.
2. **R**(va), the vertical reaction is simply the balance of vertical forces (worked out as for a simply supported beam.
3. The horizontal forces balance in the free body diagram. **R**(ha) acts at the upper part of the segment (cracks on underside assumed)
4. Failure of the stone is based on a force that equals **R**(ha).
5. From local testing of the stone, the crushing strength of the sandstone from which the bridge is built is **10 MPa** (N/mm^2).
6. The bearing area of the stone voussoir is half the total face area.
7. The vehicle weight **P** is assumed to act at the third span (or hinge) point at failure.

The aim is to find the critical point of failure; not that this is a desirable outcome, but it will provide a gauge as to how much the bridge may be loaded in confidence.

Finding the Failure Load (P)

Failure of the bridge occurs when the stone crushes at a stress or pressure of 10MPa.

Depth of stone is 450mm; width of bridge is 3000mm (3m)

Failure (thrust) = R(ha)
= 10MPa x10^{-3} x 3000 x 450/2 = **6750kN**

The answer which the engineer seeks is **'what value of loading causes failure?'**

In other words P is required.

Returning to the equation where moments are balanced about the left hand side support, in figure 2.8.8, rearranging the equation for P (weight of the vehicle):

P = (R(ha) x 3h/5 - wL^2/18) x 3/L

Now, from the previous example, 'w' is the bridge self-weight (50kN/m), 'L' is the span (30m), 'h' is the rise (10m), 'R(ha)' is 6750kN.....now solving for P:

P = (6750x3x10/5 - 50x30^2/18) x 3/30
= **3800kN** (380 tonnes)

The Bridge is Safe

The bridge is estimated to support 380 tonnes as a point load at the third span point. Given that one lorry may weigh 35 tonnes at the most and the bridge is narrow anyway, the public may rest assured that this bridge is in little danger of collapse.

Supposing the Stone was Weaker?

Supposing tests were carried out on the stone from which the bridge was constructed. It is found that the stone is weaker than previously thought. An allowable strength of, say, 1.5N/mm^2 (mPa) has been published by the testing house and sent to the highway authority.

Recalculation for Weaker Stone

Recalling the previous section on the self-weight of the bridge, the calculated stress under self-weight is 1.39 MPa (kN/m^2), which is close to the strength.

Recalculation of the failure thrust yields,
R(ha) = 1012kN.........

Recalculaton for the critical loading yields,
P......
P = (1012x3x10/5 - 50x30^2/18) x 3/30
= **360kN** (36 tonnes)

Time to close the Bridge? 36 tonnes approaches the weight of a loaded truck

Time to close the Bridge? 36 tonnes approaches the weight of a loaded truck.`

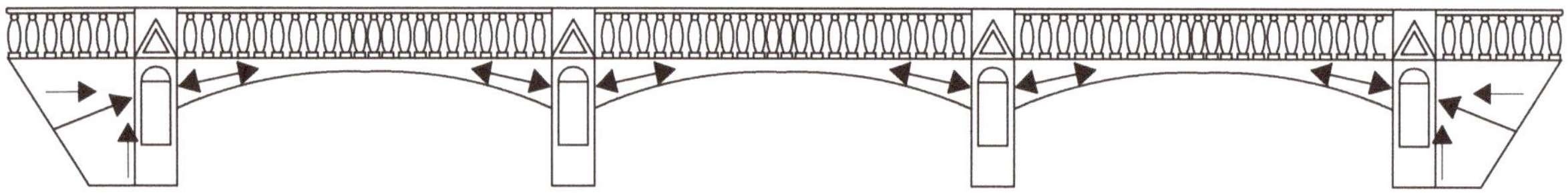

Figure 2.8.10 Multiple Arches

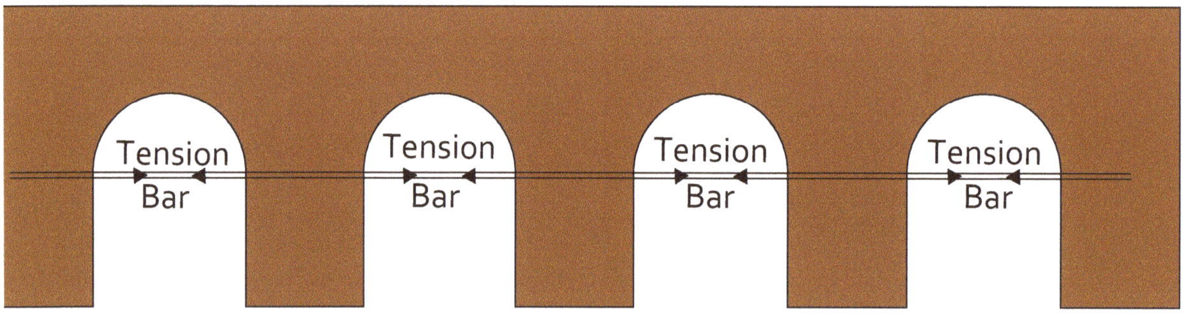

Figure 2.8.11 Multiple Tied Arches

Vaulting – Repeated Arches

Traditionally, arches are suited to multi-span structures. They were frequently employed in basements to support ground floors as a stable way of resisting the incoming pressure of ground forces and avoiding the use of timber floors, which are vulnerable to dampness and rot. Essentially the trusses are supported at the ends and intermediate supports which resist the vertical reactions. The horizontal components are resisted by adjoining trusses; being equal and opposite, they cancel each other out.

At the end bays, as illustrated in **figure 2.8.10**, the end reactions are generally buttressed with solid masonry walls, which in a basement structure are more or less absorbed by the adjoining ground, which is trying to push in on the ends.

Figure 2.8.11 shows a variation in the multiple-span arches where the horizontal forces are tied by iron bars or similar.

This method has been used to repair arches that have, in time, become unrestrained at the ends tending to push out, causing consequential movement as the internal arches play outwards. The retrofitted ties help to arrest further movement.

Floors have been supported with vaults, which tend to have a generous rise. The coliseum in Rome has concourse areas where spectators gain access to the viewing tiers display a dramatic sweep of stone vaults supported on massive columns. See **figure 2.8.12.**

Figure 2.8.13 shows a tied arch floor at the Museum of Science and Industry in Manchester. Deep wrought iron riveted girders with brick vaults carried heavy warehouse loading. The ties between the girders stabilise the lower flanges from splaying sideways. The brick vaulted floors used to be an effective way to span flooring between columns in the era before the widespread adoption of reinforced concrete.

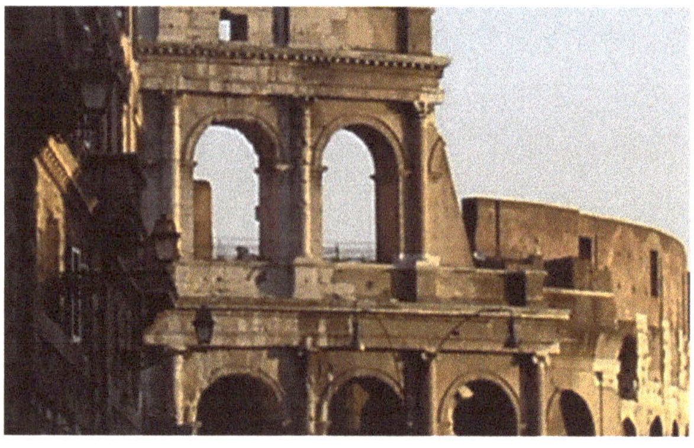

Figure 2.8.12 Vaulted Tiers - Coliseum Rome

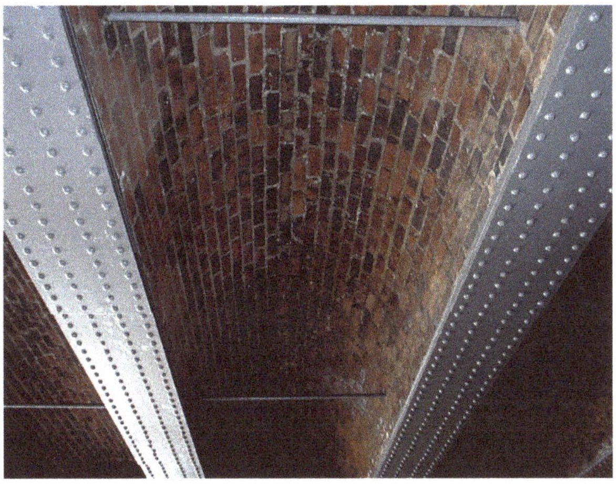

Figure 2.8.13 Tied Brick Arch Floor Manchester

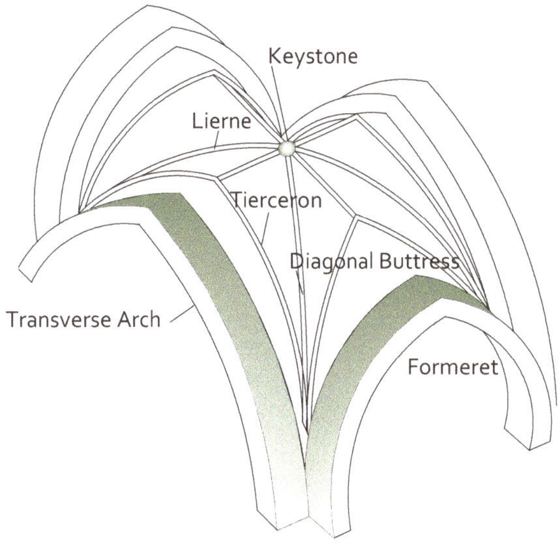

Figure 2.8.14 Anatomy of a Typical Gothic Arch

Vaulting of Medieval Cathedrals

The pinnacle of Gothic construction was the medieval cathedral. It is recommended to read 'The Stone Skeleton' (Heyman 1995, reference 5.).

Figure 2.8.14 illustrates the anatomy of typical Gothic vaulting adopted in the Middle Ages.

Supporting the vaults along the building were the transverse arches. The diagonal buttress connected the vault corners via the keystone. Tiercerons were the ribs (a stunning example at Exeter Cathedral is shown in **figure 2.8.15**).

At right angles was the formeret which spanned perpendicular to the transverse arch. In **figure 2.8.14** the formerets form the enclosure to the upper clerestory windows.

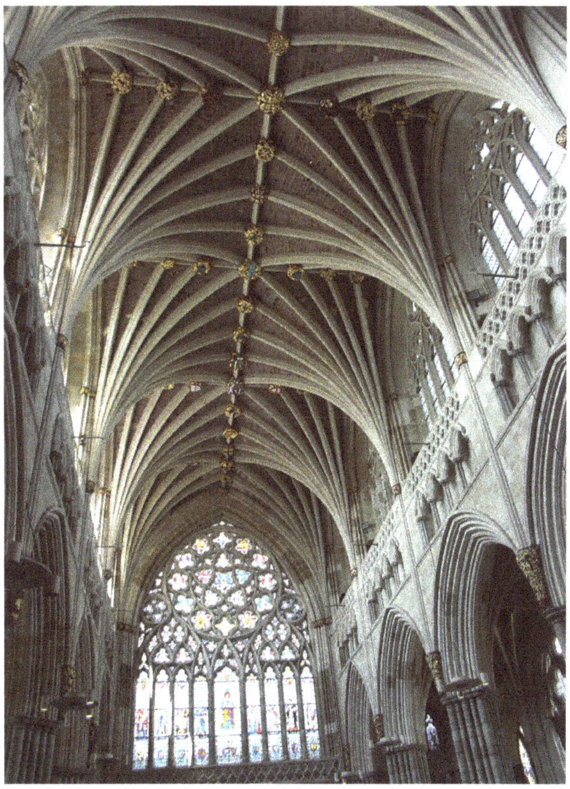

Figure 2.8.15 Exeter Cathedral Gothic Vaulting

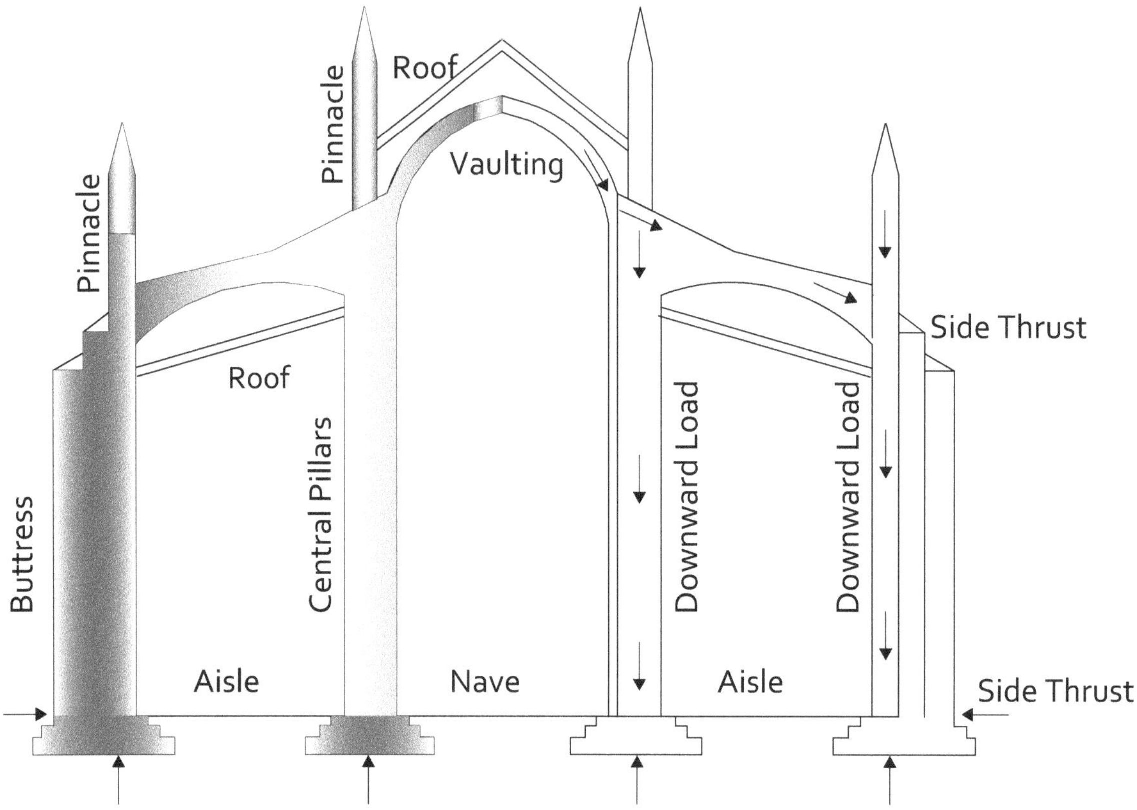

Figure 2.8.16 Typical Cathedral Section

The Buttressed Cathedral Vault

Figure 2.8.16 illustrates a typical cathedral section. The vaults are supported by central pillars, which carry their weight to the ground. Vaults exert a sideways thrust like arches. This thrust is resisted by the buttresses via the *flying buttresses* over the roof.

To support the outward splay of the vaulting external buttresses (or piers) restrain any outwards movement of the vaults.

The heavy weight of the buttresses - sometimes with heavy statutes sitting on top - is essential to resist the side thrust. Illustrated with pinnacles in **figure 2.8.16**.

Buttresses of this type are known as flying buttresses or, to put it another way buttresses with flying arms.

Should the buttress foundation start to rotate, this can have the effect of trying to stretch out the flying buttresses. Excessive movement can lead to instability and collapse of the arms and restrained vaults.

A famous catastrophic collapse of choir vaulting occurred in 1284 during the construction of the Cathedral of Saint Peter of Beauvais in France. This was thought to have been caused by the failure of slender pillars (in the location of the piers noted as 'central pillars' in **figure 2.8.16**). The slender pillars were twinned with thicker pillars internally.

These pillars supported a *tas de charge* (thrust block), which transferred the lateral vault loading onto the adjacent flying arms. It was thought that the movement of the tas de charge led to the disconnection of the flying arm.

Since the flying arm would no longer have been propping the vault, the collapse was believed to have occurred through the removal of this essential load path.

Buttresses and the Middle Third Rule.....

The question that may be asked by the observant is: what stops the buttress from toppling over?

A simplistic answer is: if the overturning moment from the horizontal thrust (multiplied by 1.5 safety factor) does not exceed the restoring moment from the total buttress self-weight, it will be stable. See 'The Stone Skeleton' (Heyman 1995, reference 5.).

> $P.D_b/2 \geq 1.5.h_{b+w}.H_b$ for stability - see calculation

A simple rule was devised from this, which is presented in the calculation which follows. If the buttress depth is divided into three equal parts, then as long as the vector resultant of *arch thrust* and *buttress weight falls* within the *middle region*, then the buttress is stable. *This does not relate to the middle third rule, for instance, presented in* **chapter 2.13, figure 2.13.14**.

With the force flow shown in **figure 2.8.17** the outward thrust of the vault has been approximated to the arch thrust discussed in previous sections. '**W**' is the total weight of the vault (wL); therefore outward thrust approximates to **WL/8h**.

The weight of the buttress is P. The force flow that finds its way to the foundation is represented in **figure 2.8.17** by a triangle of forces.

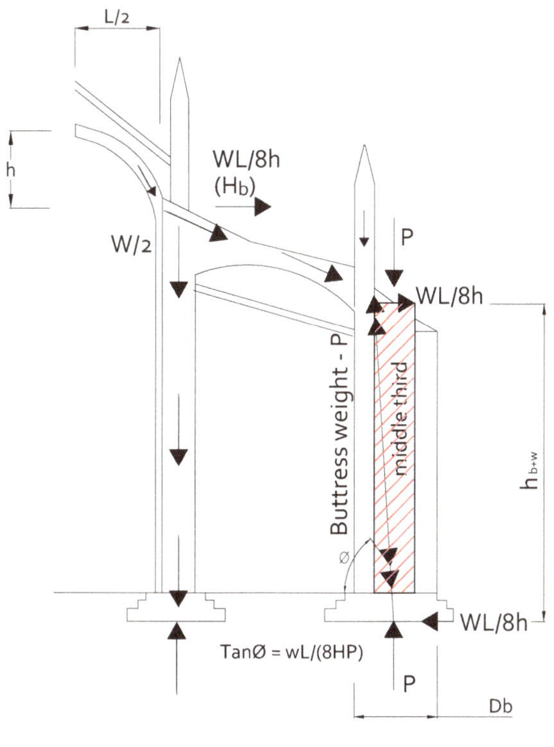

Figure 2.8.17 Buttress Force Diagram

Assuming that the buttress is a uniform width (and uniform section from the eaves level), the angle Ø may be found as in the following example.

Finding the Angle, Ø

1. Draw the buttress to scale in elevation.
2. Divide the width into 3 parts.
3. Draw a line connecting the outer points represented by the hatched section (noted previously as the middle third).
4. Calculate or measure the angle with a protractor.
5. Solving for P (weight of buttress):

Rearrange the equation:

> Tan Ø = WL/8h i.e. P = WLTan Ø/(8h)......

Example Buttress Assessment:

Span of vault, **L** = 10m
Weight of Vault, **W** = 95 tonnes (950kN)

Rise of Vault, **h** = 5m
Depth of buttress(D_b) = 4m
Middle Third Width = 4/3 = 1.33m wide

Height of buttress + footing(h_{b+w}) = 27m

Width of buttress = 2m (*measured into page*)

Weight of buttress,

P = 4x27x2x2.5 = 540 tonnes

...stone density approx 2.5 tonnes/cubic metre

From geometry, Tan Ø = 27/1.33 = 20.3...
Therefore Ø = 87°

Buttress resultant thrust (H_b) - *horizontal*
= WL/8h = 95x10/8/5 = 24 tonnes

Required buttress weight (**P**):

24tan Ø = **457 tonnes**

...less than the actual weight (above) 540 tonnes. The buttress is heavy enough on first appraisal to resist the arch horizontal thrust.

Buttress: Discounted Factors

- *The horizontal thrust is overestimated (the vault is often close to vertical at the 'tas de charge' point of bearing onto the flying buttress).*
- *The angle of the flying vault is not taken into account – which is helped by the fact that it is on a rake (or slope).*
- *The weight of the pinnacle is not taken into account. This is an ingenious method to add weight to the column. It is usually less than 10% of the pier weight its main assistance being in frictional resistance where the flying vault meets the pier.*
- *Flying buttresses are mostly arched in form. The thrust line in the arch may meet the pier closer to a vertical inclination than the assumed straight buttress in the example.*

This is a crude calculation but instructive. It illustrates the first stage of engineering judgement, in the process of design (or evaluation of an existing building). As precursor to a more sophisticated analysis it gives a good indication of the initial results anticipated.

Should the pier move sideways slightly owing to the flying buttress thrust or ground settlement, this may not be disastrous. The vaults may undergo cracking and settle, like an arch with hinges, into stable equilibrium. A glance at the underside of medieval gothic vaults confirms that this is often the case.

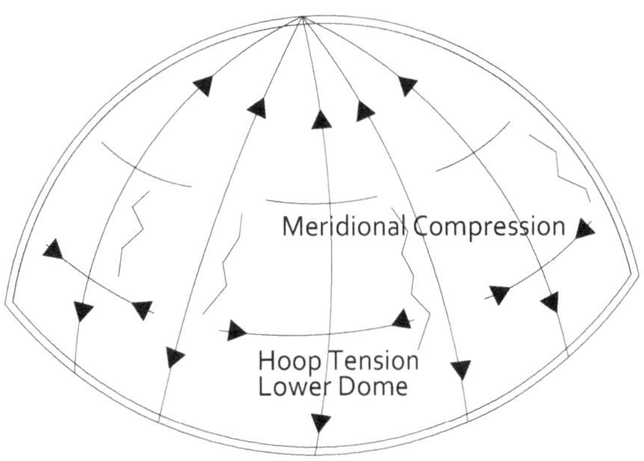

Figure 2.8.19 Semicircular Dome Section

Domes and Poleni

Some of the world-famous domes were described by Jacques Heyman, author of the Stone Skeleton (Heyman 1995, reference 5.) and various papers. He covers eloquently the subject of masonry domes (and cathedrals).

In essence, a dome can be analysed with the use of simple statics in a similar manner to arches.

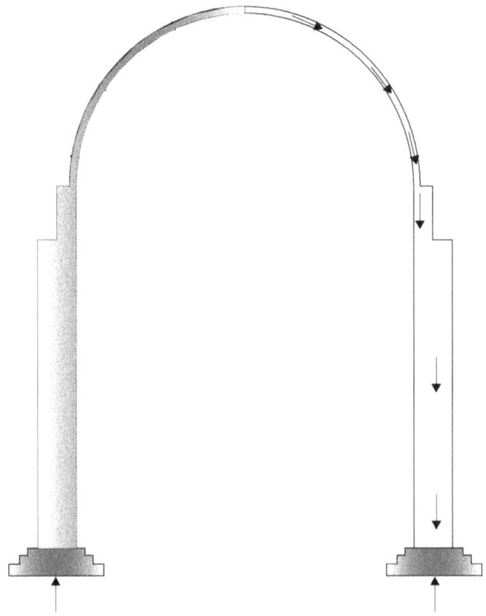

Figure 2.8.18 Dome Internal Forces

Figure 2.8.18 shows that, as a three-dimensional structure, a dome is like an arch. Ignoring any internal bending moments, it supports load by a combination of hoop stresses and meridional stresses acting at any one point perpendicular to each other. **Figure 2.8.19** slices through a dome, showing how a purely semi-circular section should not exert any lateral thrusts around its edge, with the weight carried straight down pillars to the ground.

In 1748 a polymath named Poleni analysed the dome of the Santa Maria Del Fiore in Florence by simple statics. He first sought the forces by use of a hanging chain, dividing the dome into 32 slices and hanging weights, each representing the dome segment. His first assumption was most important: that the cracking to be seen in a dome eliminates consideration of the hoop tension since masonry cannot carry the force in that direction. By this simple method, a dome may be broken into segments and analysed as an arch with hinge points.

Horizontal or Radial Thrust at the Base

Figure 2.8.19 shows a semi-circular dome. In most cases, domes are geometrically *cut-off* - i.e. the point at which they rise from the supporting pillars or arches is higher than the theoretical semi-circular *cut-off* point. In other words, the base of the dome slopes at the point of support. Therefore the arch internal compression will exert a vertical and horizontal thrust component at the point of support.

This thrust exerted at the base will depend on the *cut off* angle of the point at the support point of the dome.

An approximation of outward thrust follows in a 45° cut-off follows.

A spherical dome is shown in the section in **figure 2.8.20**.
From geometry, the span is **1.41R** and rise **0.29R** of the radius defining the sphere. The unit weight (**w**) is the density of the structure multiplied by the thickness of the shell.

Taking an approximation of the dome segment as a barrel vault with catenary shape the vertical reaction is taken as the weight evenly distributed over the spherical surface area. As the cut-off is 45°, the horizontal component is the same at 0.41wR.

This is the circumferential loading at that point, which is distributed evenly around the dome base. To obtain the load per unit length, the reactions are divided by the circumferential length (1.41πR).

Circumferential Loading

Load per unit length (vertical/horizontal) = 0.41wR

The internal compressive force
= √ ($0.41wR^2+0.41wR^2$) = 0.59wR

The smaller the cut-off angle measured from the horizontal, the smaller is the outward thrust ; tending towards zero for 0°.

The surface area of the segmented dome to obtain load is:

2.π.R.H = **$1.82R^2$**,........if H = 0.29R

Finding Horizontal Thrust and Shell Compression

Total Weight of Dome Segment = π.R.H.w

Distributed Load/unit length (vertical) ,

Vertical Reaction,
Rv = 2.π.R.H.w/2.π.R cos Ø = H.w/cos Ø

Horizontal Thrust, Rh = V tan Ø
(zero when Ø = 0)

Compression in Shell, C = √ (H^2+V^2)

Resisting the Horizontal Thrust

The cut-off produces a significant thrust that needs resisting. In later chapters, ring-beams will be reviewed, in which this example will be discussed. Having found the thrust at the base, this may be resisted in two ways:

1. Buttressing (e.g., Hagia Santa Sophia, Istanbul).
2. Ring beam or Chain (St Paul's Cathedral, London or Santa Maria Del Fiore, Florence).

Thickness of Shell

Assuming no internal bending (hinge formation or cracking within the dome), then the thickness of the masonry structure may be relatively thin. In historic structures, the thickness is around 5-10% of the span.

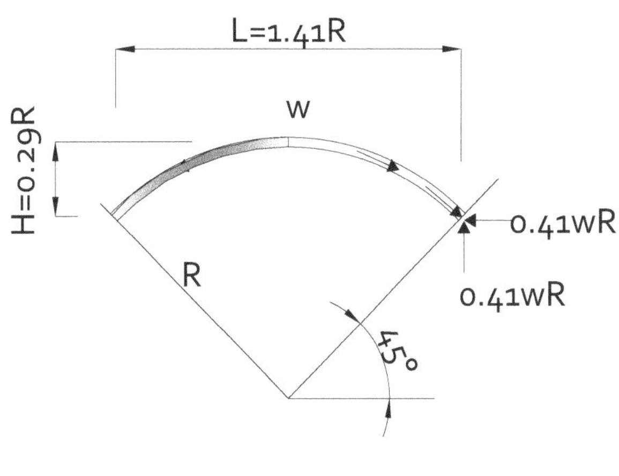

Figure 2.8.20 Dome Reactions 45 ° Cut-Off

Thin Shell Structure Example (Eladio Dieste)

Figure 2.8.22 is a representation of the Refrescos Del Norte roof designed by Eladio Dieste, a Uraguyan engineer, who created many thin-shelled structures in his native country. It is a repeated vault structure built from pre-stressed brickwork, a local material available at a relatively low cost. The shells were no more than 100-150mm thick.

The vaults were formed on timber and steel centring, and pre-stressing wires were laid on the crown and looped around the ends, pulled together to form the pre-stress. Bricks were laid over and grouted in place.

Dieste was pre-eminent in a generation of engineers who turned to mathematical methods to prescribe shapes that would minimise bending moments internally.

The Refrescos Del Norte in **figures 2.8.21** and **2.8.22** were catenary shaped repeating vaults. Dieste used stiff edge beams to transfer the end horizontal thrusts into shaped columns. At one end (see figure) there was a central column over which the roof spanned as a hybrid cantilever-vault.

Figure 2.8.22 illustrates the hybrid nature of the structure.

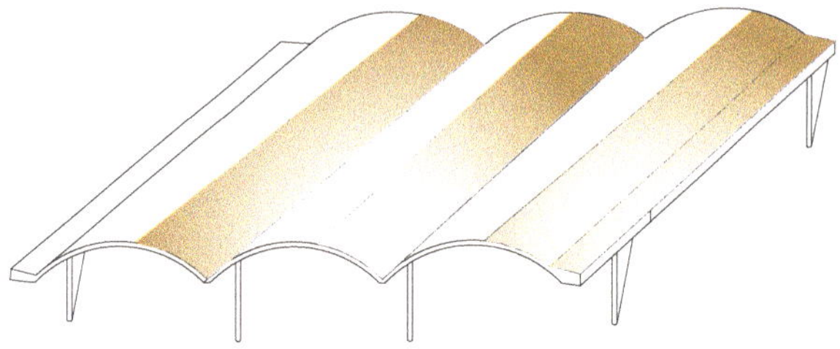

Figure 2.8.21 Refrescos Del Norte – Eladio Dieste

Three Principal Components of Structural Action

1. Arch action – multiple vaults of rise **H** and span **L**. The ratio of **L/H**, which Dieste customarily spans up to 2.5 to 4 times rise.
2. Longitudonal spanning action. The ratio of the length of the vaults, expressed as **D**, and the vault spans is such that when D is greater than 3 times the vault span (i.e., elongated in shape), the whole construction tends to act as a monocoque shell (a beam) which spans between the columns.
3. Stiffened edges. The horizontal thrust is reacted by a stiffened beam and then transferred into a rigid column.

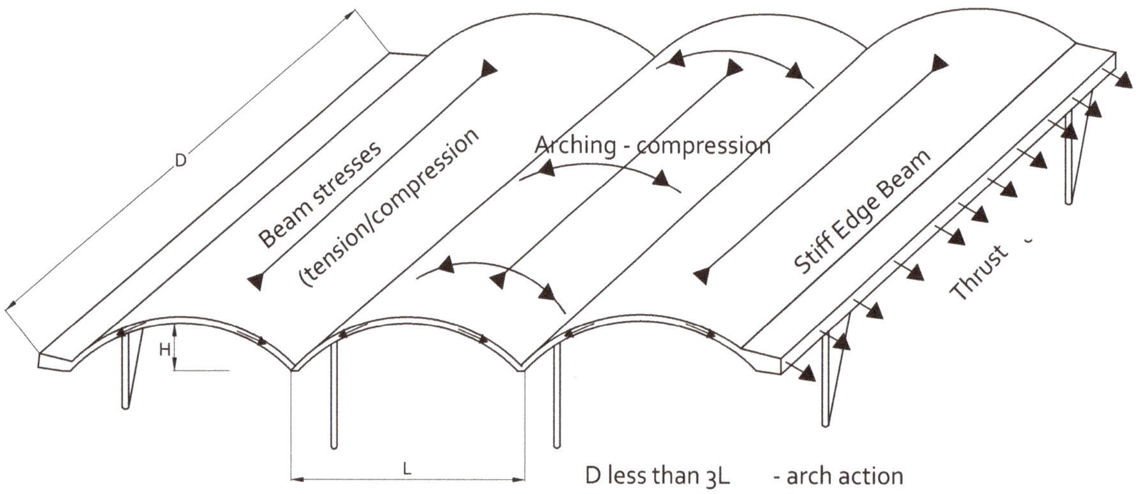

Figure 2.8.22 Refrescos Del Norte Structural Action

Hyperbolic Saddle and Thin Shell Roofs

When a structure is constructed in the shape of a hyperbola, there is essentially only tension or compression in the structure, which enables thin shell construction to be used.

A number of stunning structures using thin-shelled reinforced concrete were built in the 1950s to 1970s, after which they fell out of fashion. A good example is the John F Kennedy TWA flight center (now terminal 5) in New York. This was attributed to architect Eero Saarinen. See **figure 2.8.23**.

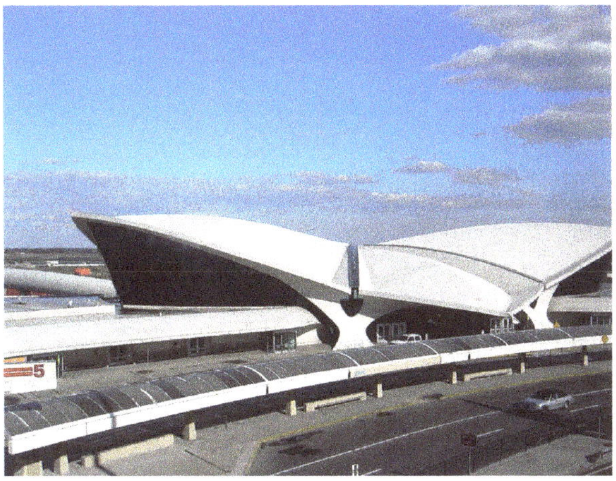

Figure 2.8.23 TWA Terminal New York

Design Of a Stadium Roof in Simple Steps

All of the chapters up to this point should have furnished the reader with enough structural principles to analyse most structures.

The remainder of this chapter will demonstrate that a stunning stadium roof can be designed with the approximate methods described hitherto. In fact, this particular roof, which may be attributed to Stephen Morley, as a structural engineer, will have been conceptually designed using very similar methods described herein.

This example is based on a real structure. See **figure 2.8.24**.

The aim is to size the tubular arch, which is dimensionally similar to the Golpharm Stadium in Huddersfield UK. The arch comprises 3 steel tubular sections in a trussed arrangement. The top two tubes are larger than the bottom tube. The roof is hung from the arch corresponding to the rear column positions illustrated in **figures 2.8.25** and **2.8.26**.

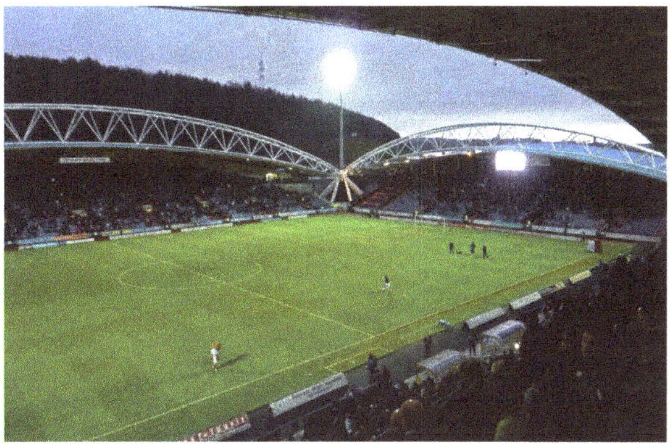

Figure 2.8.24. Golpharm Stadium Huddersfield UK

Assumptions

1. The arch under uniform load incurs no bending – i.e. is a pure arch.
2. The end supports are pins that are restrained in a horizontal direction.
3. The bottom tube in the triangular section (see **figure 2.8.26**) serves only to carry out of-balance loading (in bending – see later).
4. The top two tubes are the load-carrying arch members for a uniform load.
5. The arch rise is measured to the centre of the arch.

Dimensions and loads are shown in **figure 2.8.26**

Roof Beam Loading

First calculate the area of roof supported by the arch. Note above that the roof consists of a series of steel beams with circular holes in the webs. Roof beams are supported at hanger points.

The roof section shows that the roof beams are cantilevers with back-spans. Therefore the arch hangers will carry a higher proportion of roof area. A proportion of 60% carried by the arch and 40% by the rear columns has been assumed.

Loading (roof weight and imposed) is assumed to be a uniform load (**w**) of **2kN/m²**.

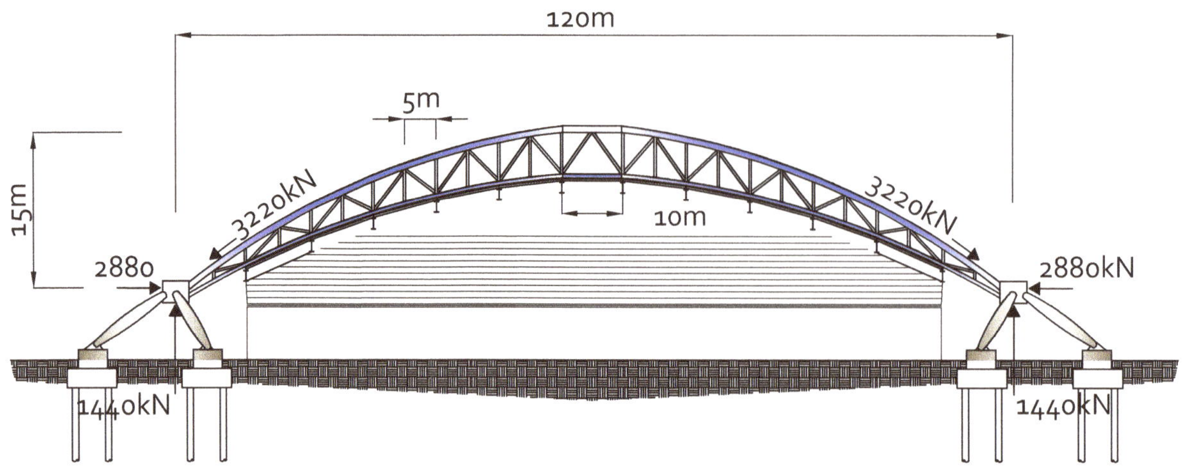

Figure 2.8.25 Stadium Elevation

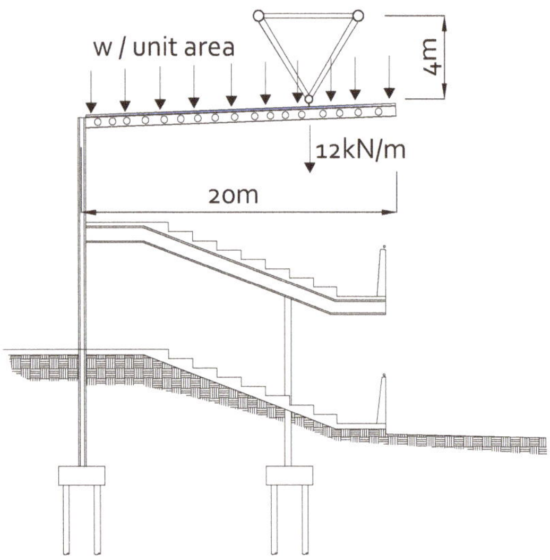

Figure 2.8.26 Stadium Section

Arch Loading

Arch loading is calculated using simplified arch formulae. The calculation procedure follows.

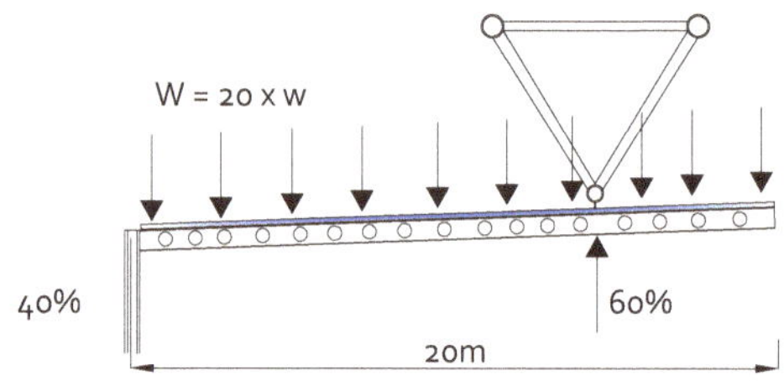

Figure 2.8.27 Roof Section Loading

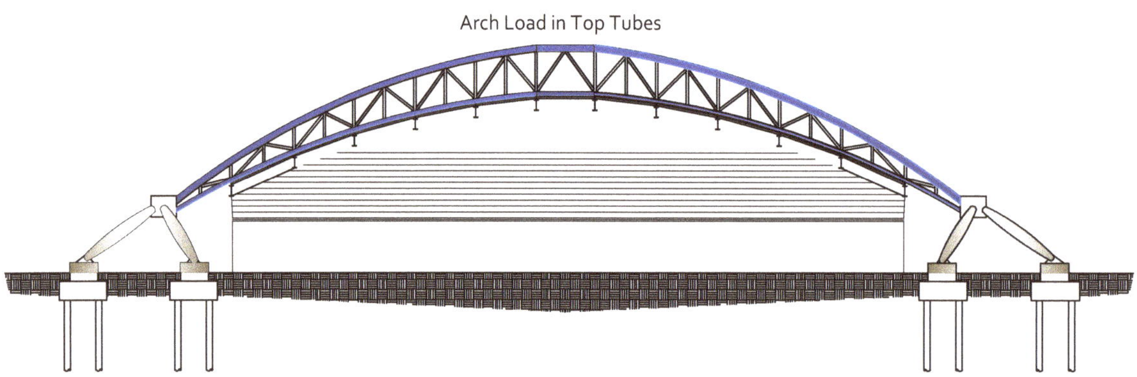

Figure 2.8.28 Top Tube Loading

Calculation of Arch Force

Arch compression (top tubes):
see **figures 2.8.25** to **2.8.27**

Approximately 60% of the roof width is carried by the arch and 40% by the columns to the back of the stand (the beam is similar to a short cantilever with back-span - see **figure 2.3.7** right hand support (3.125/(3.125+1.875).

Line load supported by arch
= 2 (roof load) x 20 x 0.60 = 24 kN/m

Reference figure 2.8.6 ($wL^2/8H$):

Horizontal Reaction
= 24kN/m x 120^2/8/15 (H) = **2880kN**

Vertical Reaction
= 24kN/m x 120 (L)/2 = **1440kN**

Note the arch is a low angle at the support, the vertical reaction is expected to be less than the horizontal reaction - just by looking at the angle of inclination. A scale sketch is of use to appreciate this.

Thrust in the arch
= $\sqrt{(2880^2 + 1440^2)}$ = **3220kN**

(i.e. angle of inclination of reactions
Ø = cos -1 (2880/3220) = 26°

- this looks correct by observation of the scale elevation of the arch. A scale sketch helps to check visually the proportion of calculated reactions and inclination.

Limit State Top Tube Loading

The loading used in the design is generally dead loading (permanent action). This is the self-weight of all of the structure and all of the finishes (roof cover and so on).

The roof dead load was taken as **1.40** kN/m² (kPa) .

Live Loading: These are the loads imposed after construction. For a roof, this amounts to snow or maintenance access.

The roof live/snow load was taken as **0.6** kN/m².

Ultimate Top Tube Loading

For the first calculation the loads were added and assumed to be uniform across the roof.

Total load (Dead and Live)
= 1.40 + 0.60 = 2.0kN/m²

In *limit state design* which is used in the UK, Europe and across many parts of the world the load is factored to include a safety margin in case of error or unanticipated effects. For the sake of this illustration a factor of 1.5 will be applied to the loads. The elements (tubes) will be selected from a Eurocode chart for buckling. Ultimate load factor to apply to the loading (and hence the calculated forces) is 1.50.

Top Tube Loading

Ultimate Tube load in the top 2 tubes assuming they resist all of the arch thrust is:

Ultimate Tube Load (shared by two)

= 1.5 (factor) x 3220 (arch thrust) / 2
= **2415kN**

Sizing the Tubes

Compression elements are designed by assessing the buckling load. A designer will generally refer to printed tables to save time (or use the computer software to select the element). The tube size may be selected from a buckling table, which may be obtained for most codes.

Euler (**figure 2.7.23**) may be used, but using a table avoids trial and error calculations since a further check for compressive yield stress (**P/A**) would need to be considered, which may be less than the Euler value. Also, Euler values are not as safe as modern formulae based on Perry-Robertson formulae.

See **figure 2.8.29**.

Effective Length

A compression element has a given failure or critical load for a given length. In the arch-truss, in this example, the cross struts stop the top and bottom tubes from intermediate buckling.

A buckling table is illustrated in **figure 2.8.29**.

Critical buckling length (effective length, Le)
– *distance between cross struts*

Le = 5m (**figure 2.8.25**)

Selection of Top Tube

The designer now has to hand:

1. Tube load - 2415kN.
2. Effective length between cross restraints – 5m.

The tube size selected from **figure 2.8.30** is a 273mm diameter tube with 16mm wall thickness.

This has a critical failure load of **2576kN** at 5m effective length (i.e. greater than **2415kN**).

Circular Hollow Sections in accordance with EN 10210 - UK Edition to Axial

Designation		Grade S275 (Design Grade 43)						
Size D	Thickness t		Compression Resistance in kN for Effective Length in M					
mm	mm	Class	2	3	4	5	7	8
193.7	6	1				571		
219.1	16	1				1759		
244.5	5	1				722		
273	12	1				1980		
273	12.5	1				2050		
273	16	1				2576		
323.9	5	1				1095		
323.9	6	1				1306		
323.9	6.3	1				1371		

Figure 2.8.29 Tube Critical Load Chart

Selection of Bottom Tubes

As noted before, the bottom tube resists any *out-of-balance loading.*

This generally refers to live loading. For instance:

- Snow drifting onto the tip of the roof. This is not a problem if the blanket snow load used before is greater than the peak loading on the tip.
- Snow on the part of the roof. Say heaters below on the terrace melting the outer parts and leaving the middle (and reverse of this).
- Wind uplift – wind suction on a lightweight roof can cause a net uplift effect.

The art of selecting load cases is to avoid going to extremes. This can seriously affect the design for no particular reason other than the engineer being over cautious. On the other hand lack of serious consideration of worst-case conditions is negligent and could lead to structural failure.

In the first decade of the 21st century there have been a number of roof failures under snow loading: a swimming pool roof in Moscow in 2004, collapse of a roof over an ice rink in Bavaria in 2006, killing 15 people, and a supermarket roof in Germany in 2006.

Considering snow in the middle only, as discussed in chapter 2.2 for the simply supported beam, a very approximate case will assume that 1/8 of the snow is crowded onto the middle of the roof. See **figure 2.8.30**. 2.8.30. Since the dead loading is uniform, the live load only needs be considered for this *out of balance effect.*

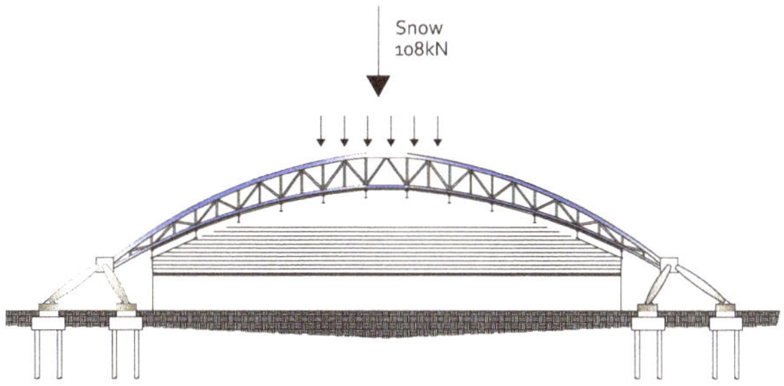

Figure 2.8.30 Concentrated Snow Load

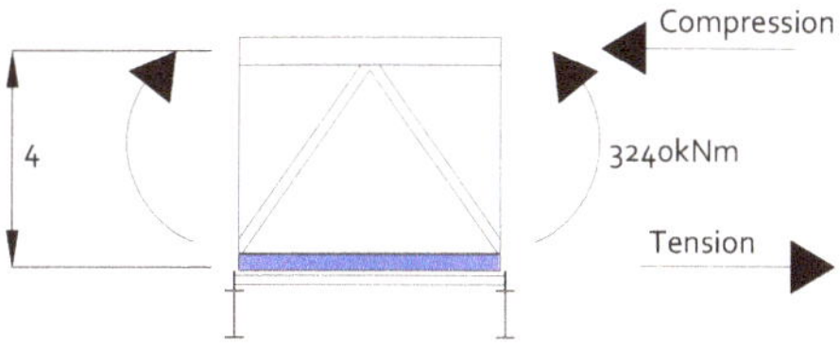

Figure 2.8.31 Resolution of Bending - Chords

Check Out of Balance Bending

Calculating the Bending Moment Due To Asymmetrical Snow (**figure 2.8.31**)

Returning to chapter 2.2 an even *rougher* approximation for bending assessment will be that the arch is straightened out into a simply supported beam and that the snow load is represented by a concentrated point load in the middle of the roof.

Central point load (snow at 0.60 kN/m²; ultimate factor 1.50):

Calculate idealised snow point loads

P = 0.60 (60% factor) x 20 (roof width) x 0.60 (snow) x 120 (span)/8 = **108kN**

Therefore bending moment (midspan)
= 108 (kN) x 120 (L) / 4 = **3240kNm**

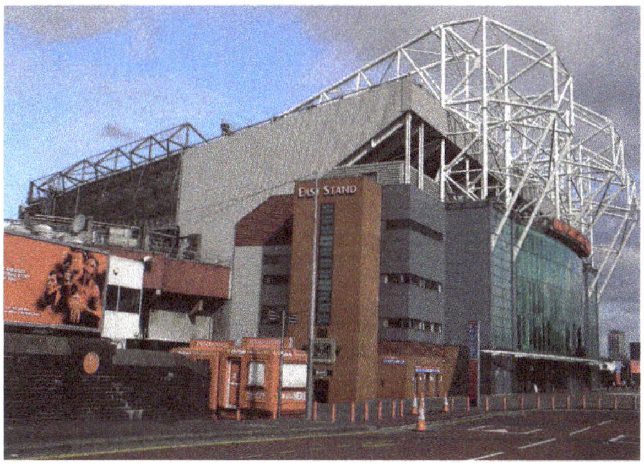

Figure 2.8.32 Manchester United Cantilever Roof Trusses

Tube Sizing

Now, the bottom tube may be sized. The arch is really a truss in section in the centre. Using the methods of **figure 2.6.8** (shown above), the compression and tension may be found in the bottom and top chords

Size From Tube Tension:

Tension in the bottom
= 3240 (M)/4 = 810 kN (T) ....

Or conversely compression in each top tube is 810/2 = 405 kN.....

The bottom tube is in tension, therefore, the critical buckling need not be considered (as in compression loading).The size is then selected from a stress area calculation or tables. A size appropriate for the area was found to be a **193.7x8** Circular hollow Section (**193.7** outer diameter and **8**mm wall thickness). This was selected from a table. A larger tube the same size as the top tubes might be chosen for aesthetic reasons.

Golpharm and Chelsea– Structural Efficiency

It is interesting to note Golpharm and Chelsea (**figure 2.7.8**) stadium roofs, both attributable to Steven Morley, do not follow the more familiar cantilever style. See, for instance, Old Trafford Stadium, Manchester United in **figure 2.8.32**.

It can be argued that the arch/suspension roof structures spanning across from end to end are more efficient than adopting cantilever solutions. The combination of efficacy, visual balance, and lightness of form in both stadia are undoubtedly graceful forms.

Chapter 2.8
Quick Note Summary

1. Arches are a stable form of construction in traditional masonry – relying on compression-only between structural elements.

2. Masonry/stone arch components are generally stressed to within low levels during use.

3. A masonry arch fails when the thrust line falls outside the middle third of the voussoirs which are (in general) the principle load bearing elements.

4. The horizontal and vertical reactions may be estimated readily and resolved to find arch thrusts –$R(H) = wL^2/8H$, (H is rise), $R(V)=wL/2$, L is span

5. Assuming a critical point, an arch may be simply analysed by assuming 3 pins. A statically determinate system.

6. Repeated arches or *vaulting* is a common construction method. The arch at the very end will need buttressing to resist the thrust. Alternatively thrust may be resisted internally within each arch using tie bars.

7. Medieval ecclesiastical buildings used the vault form to a great extent. External flying buttresses were the most common method to resist the arch thrusts. These were generally sized by the *rule of thumb*, handed down over generations, which could achieve balance – defined as keeping the component of thrust within the middle third of the buttress pier.

8. Masonry domes may be demonstrated to work using *plastic* methods – ignoring stresses in the hoop direction and resolving the meridional stresses that are in compression.

9. Domes usually need restraining at the base – using ring beams (or chains) or buttressing.

10. Thin shell structures are shaped as hyperbolas or catenaries to avoid *out-of-plane* bending forces – and hence are subject only to internal compression or tension.

11. Long-span roof structures are generally more efficient as arch or catenary forms rather than using cantilevered forms.

Chapter 2.9

2.9 Torsion Structures and Ring Beams

Torsion structures and ring beams are not common to everyday structural engineering in buildings. When loads are applied eccentrically to structures causing twist special attention needs to be taken. It is more common to design structures to avoid torsion than to check sections for torsional stresses. This chapter begins with the basics: the fundamental torsion equations for shear stress and rotation. Closed hollow sections are compared with open sections, such as channels, for resisting torsion. A practical balcony handrail problem illustrates how torsion and bending are often combined; examining the issues of relative stiffness and how torsion might be 'designed out'. Radial or circular beams are reviewed, showing how to design out torsion or to put it to good use. Finally the principle of ring beams is examined with examples: restraining the outward thrust from domes and in a basement structure capping beam around the top of perimeter piles.

Torsion

Torsion is analogous to bending. It is the application of an eccentric or offset load to a structure which causes a twisting effect.

The tube in **figure 2.9.1**. is subjected to force P on an extended arm, at a distance La from the centre of twist of the tube.

Torsion (or torque),	**T=La x P**
....expressed in SI units as kNm	

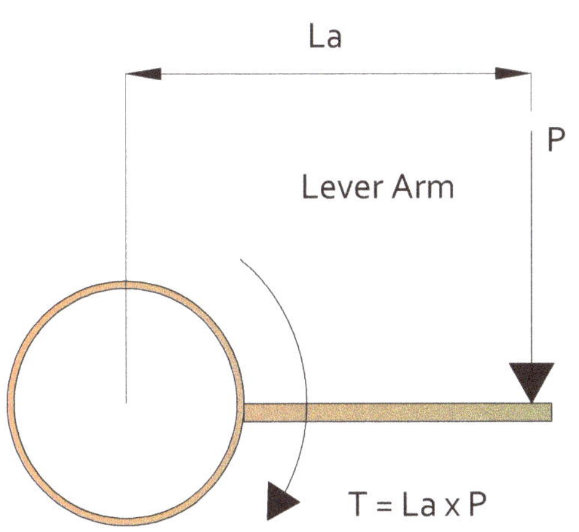

Figure 2.9.1 Torsion

In general torsional loading can be a challenge for an engineer in part because the most commonly used structural sections shown in **figure 2.9.2**. in steel design are not very effective at dealing with torsion. Whereas closed sections in **figure 2.9.3** are stiff and strong in torsion.

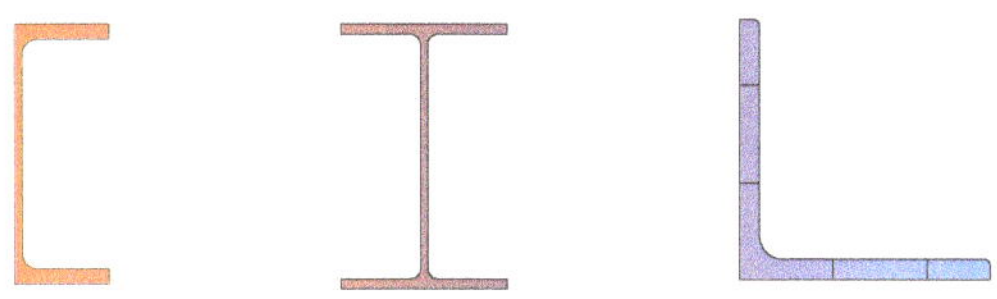

Figure 2.9.2 Open Sections

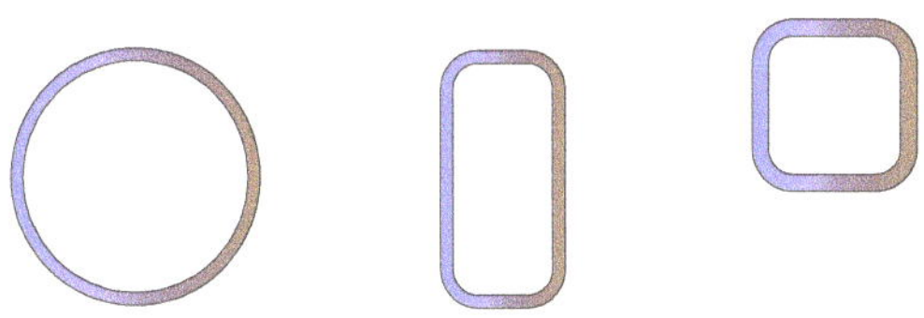

Figure 2.9.3 Closed Sections

In general a designer faced with a torsion problem will opt for either of two approaches:

1. Select a closed section.Steel closed sections are usually more expensive than the equivalent weight of an open section. In addition bolted connections involve more care to avoid *blind fixings*.

2. Use an open section but design out the torsion. For instance by fixing cross members with moment connections to resist the torsional forces applied.

Torsion Arm

The simple example in **figure 2.9.4** is a torsion arm, which is a cantilever beam of length L. In order to resist the applied torsion the beam has to have a built in support which, like a cantilevered beam (**chapter 2.3**), will not allow rotation.

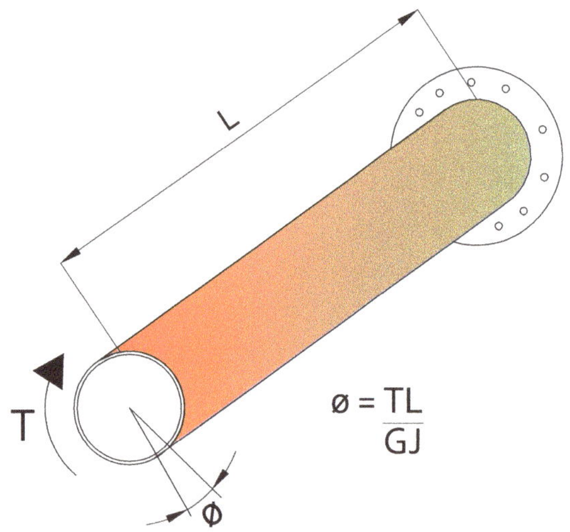

Figure 2.9.4 Torsion Beam

Torsion Rotation Calculation

From **figure 2.9.4** torsional rotations may be readily found.

> **Torsion Calculation - Rotation**
>
> The torsional rotation of a beam, with a fixed end, length:
>
> Rotation, **Ø** = **TL/GJ** (radians)
>
> The parameters: G (Shear Modulus), is related to the Flexural Modulus E by Poisson's ratio (v).....
>
> Shear Modulus, **G** = **E/(2(1+v))**
>
> J (sometimes expressed as K) is the torsional constant, which, like moment of inertia (I) is measure as dim^4.
>
> For instance, in SI units, J is generally quoted in cm^4.

Torsion Stress Calculation

Torsional induces shear stresses in a section. If torsion and bending are combined in a section then it may be necessary to check the principal stresses. A combined check may be readily found in software which outputs *Von Mise* derived stresses combining shear and axial or bending stresses.

> **Torsion Calculation - Stress**
>
> Torsional Stress is a shear stress, which is analogous to bending.
>
> A torsional modulus, C, is expressed in cm^3..
>
> Torsional Stress, **T** = **T/C**....N/mm^2
>
> The torsional modulus for a closed section tube (or square or rectangular section) is expressed below.

Note rotation is calculated in radians. A quick and simple conversion to degrees is to multiply radians by 57.

> **Example**
>
> A 100x100x10 Square hollow section with a free end (2m long) is subjected to a torsion of **10kNm**.
>
> From tables,
>
> J=761cm^4 (7.61x10^{6}mm^4)
> C=133cm^3 (133x10^{3}mm^3),
> G (steel) =78x10^{3}N/mm^2
>
> **Calculate Rotation (TL/GJ)....**
>
> Ø = 10x10^{6}Nmmx2000mm/78x10^3/7.61x10^{6}mm^4
>
> = 0.034 radians (0.034x57 = 2.0°)
>
> **Calculate Torsion Stress**
>
> T (stress) = T/C = 10x10^{6}Nmm/133x10^{3}mm^3
>
> = **75N/mm^2**
>
> (Capacity of mild steel is circa 157N/mm^2)

Closed and Open Sections Compared

A simple comparison to demonstrate the superiority of closed sections in torsion is to compare three sections. See **figure 2.9.5**.

Three sections of the same material with almost identical cross sectional area, and hence weight, for a given material are shown. The closed section (the tube), has a **J** value of **2640 cm³** which is some 240 times greater than the channel section and 377 times greater than the I section.

Comparing the moments of Inertia is instructive: the open wall I and channel sections are around 2 times stiffer in bending or flexure.

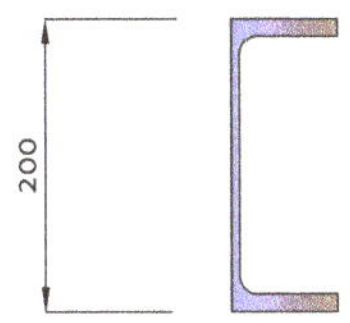

J= 11cm^4
I= 1963cm^4

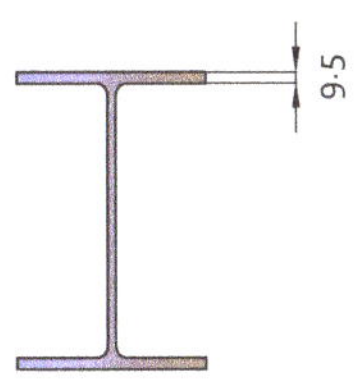

J= 7 cm^4
I= 2105cm^4

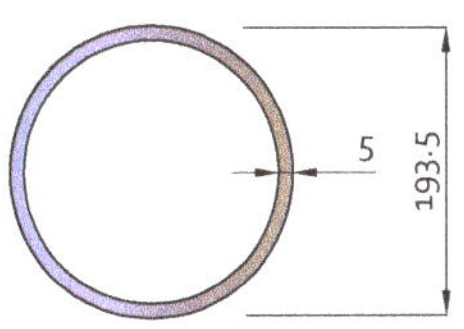

J= 2640cm^4
I= 1320cm^4

Figure 2.9.5 Open and Closed Wall Sections Compared (all 23kg/m)

Open Sections Warping

For an *open walled* section, such as an I-beam, channel or angle section, there is another complication with affects the rotation and stresses: the *warping of the walls*. The effect of warping is dealt with using St Venant methods (the St Venant effect was named after a French professor at the Ecole Des Ponts et Chaussers who presented his most famous memoirs on torsion in 1855-56).

Torsion and Bending Combined

A common problem to resolve is deflection of balustrades on balconies. Handrails have to meet the criteria of strength and deflection. A flexible handrail is off putting to the user. It may be strong enough to hold the loading imposed by a person leaning against it but if it moves easily the user is likely to perceive that it is unsafe.

Leaning on the handrail will put a lateral loading on the *balusters* (the posts which hold up the handrail). The balusters act as vertical cantilevers which like the torsion arm in **figure 2.9.1** will try to twist the supporting balcony structure.

An engineer may be called upon to size the baluster and hand railing. This cannot be done without including the balcony structure. An analysis was carried out with a side load of **0.74kN** (74kg – the weight of a person) loading the middle baluster. The balcony length is 4m long and the baluster is fixed to the top of the channel.

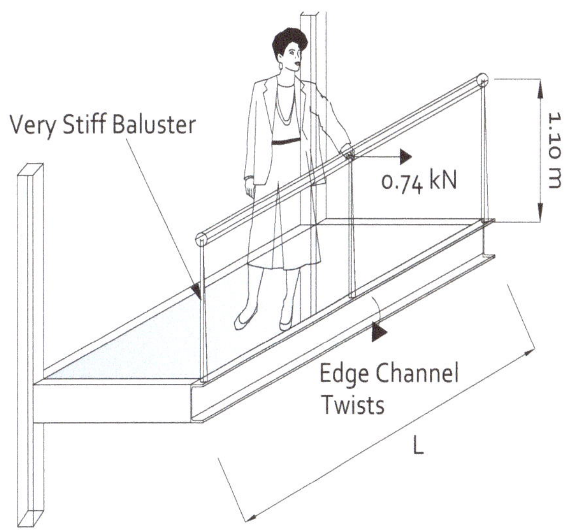

Figure 2.9.6 Balcony Torsion Problem

An edge channel size of 200x75mm was selected which is to be found in **figure 2.9.5**. Torsion is induced in the channel if the handrail is pushed. The torsion rotation was calculated assuming fixed ends. It was first necessary to calculate the applied torsion. This is equal to the bending applied to the middle baluster

Calculation of Torsion at Base of Baluster

(See **figure 2.9.6**)

Height of baluster = 1.1m high

Torsion applied to edge channel
= 1.1 (height) x0.74 (kN)
= 0.80kNm

Torsion within edge channel
= 0.80/2 = **0.40 kNm**

Calculation of Deflection at Top of Baluster

Using the torsion obtained from the previous calculation, assuming a very stiff baluster, deflection may be estimated by calculating the rotation at of the edge channel.

Rotation of baluster

Assume L = 2m (See **figure 2.9.8** for calculation)

(Ø) = TL/GJ (over half the length)
= 0.40×10^6 (T in Nmm) x 2000 / 78×10^3 (G steel N/mm^2) / 11×10^4 (J ref. **fig 2.9.4**)/2

= $0.40 \times 10^6 \times 2000/78 \times 10^3/11 \times 10^4/2$

= **0.045** radians (2.5 degrees)

Deflection at the top of the baluster may be obtained from geometry.

Deflection Top of Baluster

The movement at the top is the height x tan (rotation angle, or Ø), or h x Ø (in radians). See **figure 2.9.7**.

Deflection at top:

= 0.045 (radians) x 1100 (h in mm)
= **50mm**

- a large movement

Solutions to Stiffen the Baluster

50mm is a high deflection. The handrail when pushed will feel very flexible. Two possible solutions follow.

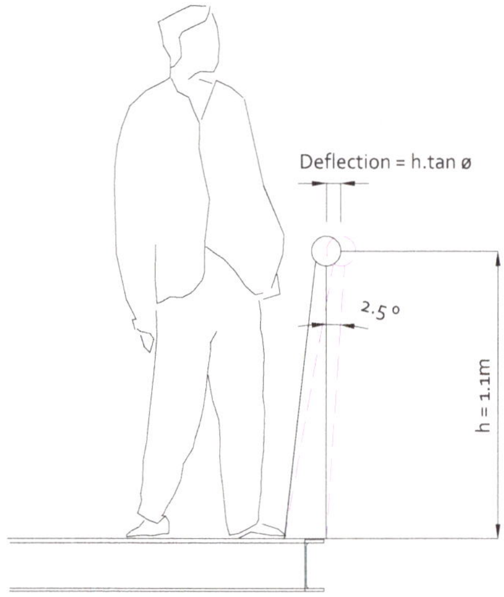

Figure 2.9.7 Baluster Deflection

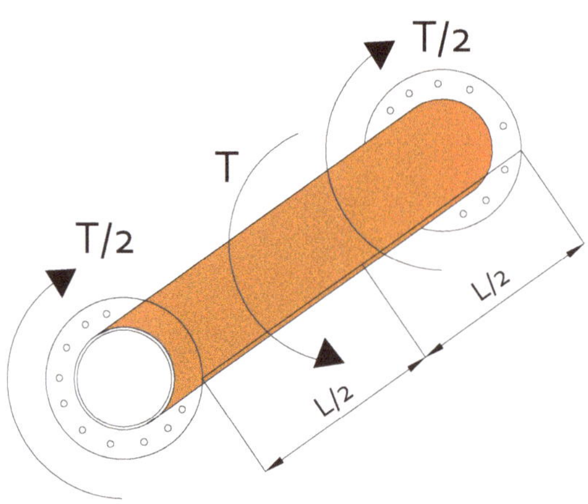

Figure 2.9.8. Torsion Applied Midspan

Solution 1

Use a Hollow Section for the Edge Beam
- 200x100x6 RHS

J (200x100x6 rectangular hollow section)
=1473 cm^4 (134xJ for 200x75 channel)

Rotation of baluster (Ø) See **figure 2.9.8** where the Length is L/2 and Torsion T/2 for a load applied midspan (imagine the tube is uct in the middle and equal an opposite torsion T/2 is applied to each end - **figure 2.9.4** applies):

= 0.40×10^6 (Nmm)x2000 (mm)/78×10^3(G)/ 1473×10^4 (J)/2
=$0.40 \times 10^6 \times 2000/78 \times 10^3/1473 \times 10^4/2$
= 0.00035 radians (0.02 degrees)

Deflection at top

= 0.00035 (radians) x 1100 (h in mm)
= **0.4mm**

Significantly reduced from 50mm.

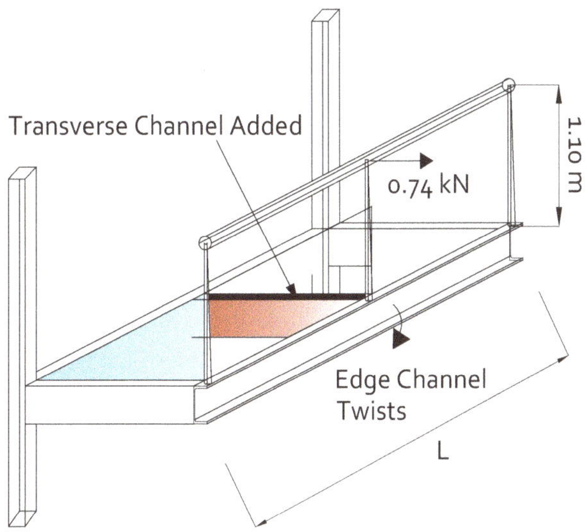

Figure 2.9.9 Adding Transverse Channel

Solution 2 (See figure 2.9.9)

Place another channel behind the baluster

A transverse channel is placed up against the baluster with a fixed connection (to transfer the torsion). The channel is a 200x75 channel (ref. **figure 2.9.5**) with I value 1963 cm^4.

Assumption: the transverse channel absorbs all of the applied torsion as and end bending moment (fixed connection) thereby taking over the role of restraining the upright from the edge channel.

The rotation (Ø) is obtained assuming a beam with a moment at one end and pin at the other.

A moment is applied to the end of the transverse channel.

Length of transverse channel (width of balcony) = 1m

Moment applied to the channel
= 0.74x1.1 = 0.80 kNm

Rotation of transverse beam (hence base of baluster) - ref. **figure 2.4.4**.....

Rotation (Ø) = ML/3EI

=(M) 0.80×10^6 (in Nmm) x (L) 1000 (mm) / 3 210×10^3 (E, steel) / (I) 1963×10^4
= $0.80 \times 10^6 \times 1000/3/210 \times 10^3/1963 \times 10^4$
= **0.00007** rads

This is very low. It equates to:

Top deflection (Hx Ø)
= 0.00007x1100
= **0.07**mm

The beam has stiffened significantly the balustrade - even more so than the previous example substituting a hollow section.

In summary

- In summary, solution 2 is more efficient than solution 1 (with the RHS closed box section); the comparable weights being similar (**54kg** for the **RHS** and **69kg** for the combined channels).
- It may be architecturally desirable to have an expressed edge channel. Another means of stiffening is needed.
- The torsional problem is designed out by strategic placement of a fixed end beam.

Aside

Relative Stiffness -A structural system with multiple redundancies.

Referring to the previous example addition of a handrail to the analysis adds considerable stiffness to the system. If the handrail shown to the left in figure 2.9.9 returns to a fixing into the wall it will assist in resisting the load in bending (point load midspan).

There are three alternative load paths to resist the force:

1. ***Handrail acting as a lateral beam restraint.***
2. ***The transverse beam under in bending.***
3. ***The edge channel in torsion.***

The total deflection at the top will result as a combined action of the three load paths.

The fraction of the applied load that each particular structural element resists is proportional to the relative stiffness of that element.

For instance if the relative stiffnesses (reciprocal ratio of each calculated deflection over the sum of the reciprocals) are:

Handrail 30%
Upright and beam (bending) 60%
Channel in torsion 10%

Then approximately the proportion of loading resisted by the handrail will be 30%, and so on. Or to put it another way, the channel is so much less rigid than the handrail its loading will be small.

Cavity Wall Torsion Beam

In the UK domestic and commercial construction employ cavity walls – with concrete blocks on the inner wall and brickwork facing on the outer wall.

When a wide opening or change, say shop space, is required under the first floor the outer leaf needs supporting. A widely used detail is shown in **figure 2.9.10**.

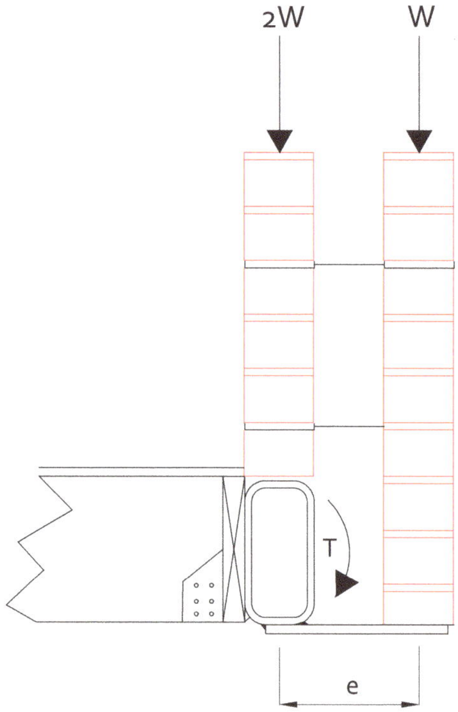

Figure 2.9.10 Cavity Wall Beam

A rectangular hollow section (RHS) beam is shown spanning between two support points with total loading summed from the floor spanning onto it from the inner wall and outer wall. It is to be noted that the supports, posts or otherwise, should be designed to resist the support torsion - which is transferred as a bending moment in a similar manner to **figures 2.3.8** and **2.3.9**.

Torsion Due to Eccentricity

The outer wall sits with an eccentricity (e) away from the centreline of the beam. Therefore there is a torsion which is applied to the beam:

Torsion due to eccentricity, $\mathbf{T = W \times e}$

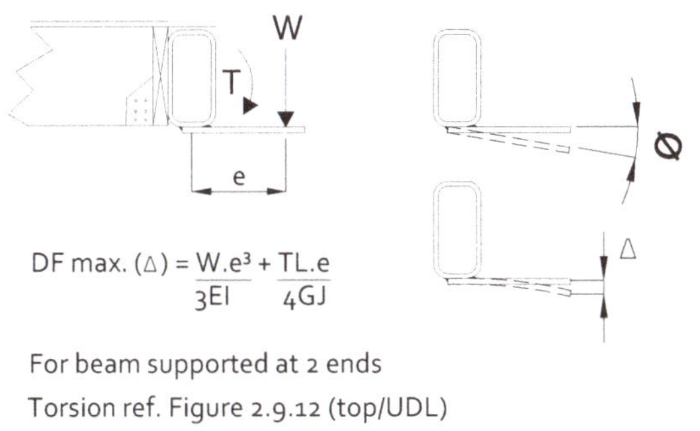

Figure 2.9.11 Local Rotation and Deflection

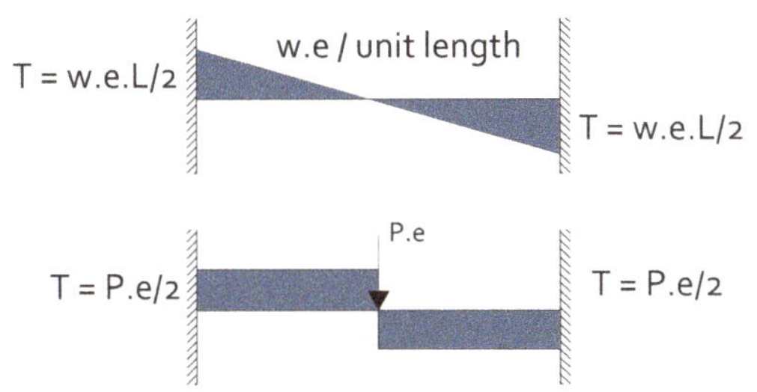

Figure 2.9.12 Torsion Distribution Beam with UDL Eccentricity

Cavity Wall Support: Check for Twisting

An essential check is the distortion of the plate supporting the outer wall – this needs to be limited.

This involves a local check. See **figure 2.9.11**.There are two components of tip deflection of the plate downwards:

1. Movement from rotation of the beam (see above)- this is calculated by working out the rotation of the beam (TL/(4)GJ for beam 2 end supports) – see **figure 2.9.11** and then this value, in radians is multiplied by the eccentric length (e).
2. Bending in the plate. The plate is viewed as a cantilever with fixed support. Returning to **figure 2.3.2** with point loading, the tip deflection is readily calculated.
3. The rotation at the supports due to the applied end torsion as a bending moment.

The moment of inertia (**I**) is that of the plate – **wt³/12** (width of the plate x thickness t^3 /12)

Note: the loading on the edge of the plate is a UDL, therefore the torsion in the beam is We.l/2 (maximum). See **figure 2.9.12** for torsion distribution in a beam.

A Curiosity – Cladding Connected to a Column with No Eccentric Moment

This particular problem is a simple resolution of forces and moments but which is not immediately obvious. The solution is indebted to Professor H.G. Allen of Southampton University who gave this problem to various students. Not one was reported to have solved it.

Wall panels or columns are fixed alongside the masonry or cladding fixed at regular centres with ties.

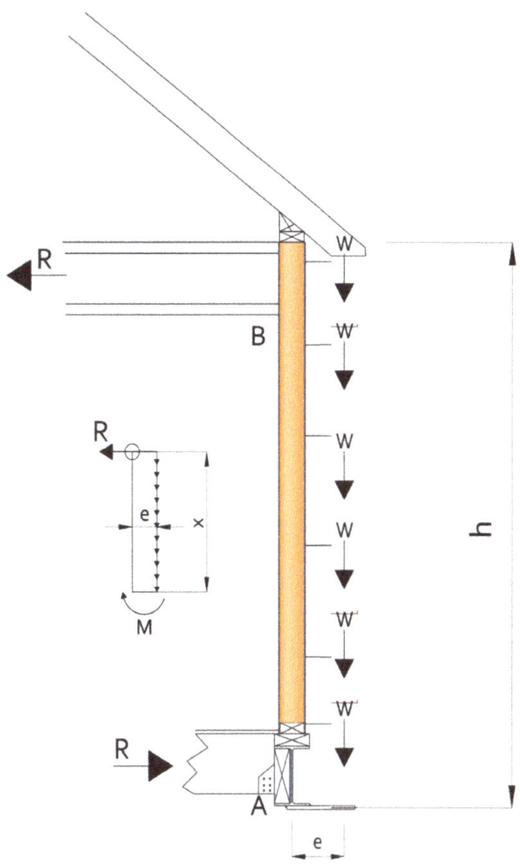

Resolving Moments about A
R.h = W.h.e therefore....R = w.e.
Resolving Moments about X in figure above
w.x.e + M = R.X
Hence....M = (R - w.e).X = 0, since R=w.e......

Figure 2.9.13 Zero Moment on a Wall or Column Intimately Connected to Cladding

Under normal circumstances it is assumed that the cladding weight eccentricity causes a torsion in the supporting beam.

However the proof offered in **figure 2.9.13** shows that there is that the moment transferred into the supporting wall and beam under is zero.

Instead the overall eccentricity is resisted by the *push and pull* forces shown as Rh in **figure 2.9.13**.

The proof is applicable on the basis that:

1. The outer skin is intimately attached to the inner wall but not supported. In this instance the outer wall is tied to the inner wall with wire ties – a common requirement for brick outer leafs. The inner wall has flexural rigidity.

2. The bottom beam is connected to the first floor in such a way that the force may be transferred into the floor. See **figure 2.9.13**.

Aside

Lateral Torsional Buckling

***Figure 2.9.14** illustrates a very important phenomenon when considering the bending capacity of beams. This relates particularly to thin walled beams but is also a consideration for timber joists.*

Lateral torsional buckling occurs when the flange of a beam in flexure, which is in compression, tend to warp or distort between restraints. This is a compression buckling which is similar to the buckling of columns (as discussed in chapter 2.7 on the aside with respect to Euler).

This is not usually a consideration for a continuous beam in the 'hogging' support moment condition.

The procedure for calculating the bending capacity of any beam is:

*Select worst case effective length between torsional restraints (Le in **figure 2.9.14**).*

Consult tables with give the buckling moment capacity of that beam. The more slender the beam (the narrower and thinner the flanges are), the lower is the buckling capacity.

The buckling capacity may also be calculated mathematically using the slenderness (Le/ry) ratio of effective length to radius of gyration.

Timber joists traditionally are fixed with timber blocking (or noggins) or with herringbone struts in centre span if they are slender to stop buckling. If the ratio of depth to width of a timber joist is no greater than 3 and the floor deck is nailed onto the top of the beam this is not usually a concern.

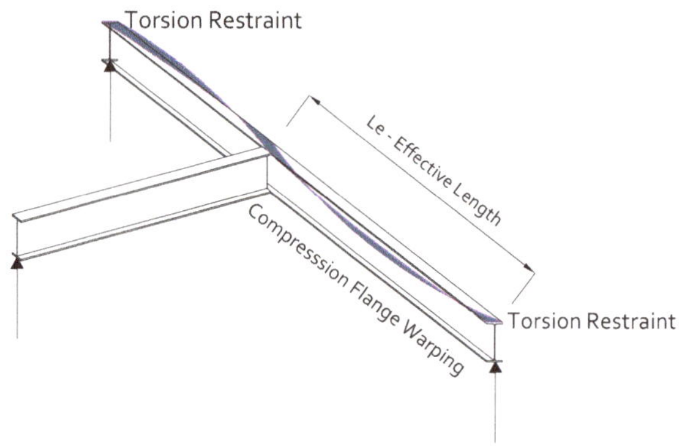

Figure 2.9.14 Lateral Torsional Buckling

Curved Beam with Rotationally Fixed Ends

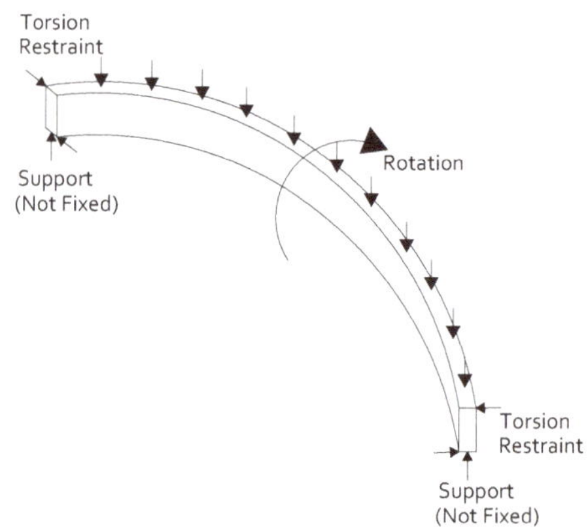

Figure 2.9.15 Curved Beam With Rotation End Restraints

The curved beam in **figure 2.9.15** has torsional restraints at the supports. The beam is also supported vertically but the ends can rotate with respect to the bending in the beam.

In other words it is pinned with respect to bending moments at the ends. **Figure 2.9.16** shows an illustrated example of where this kind of beam might be adopted. In a steel framed building a curved corner beam supports some of the floor and cladding to the outside of the building.

Is a Fixed Connection Needed?

To form a fixed connection at the two stanchions (or columns) for the curved beam may be problematical: fixed connections usually entail thick steel connecting plates and a lot of bolts to achieve the moment connection. Designing a connection to resist local torsion moments only may reduce the plate thickness and bolts numbers needed.

This problem is a good exercise checking the direction of axes in computer models; always a source of potential confusion.

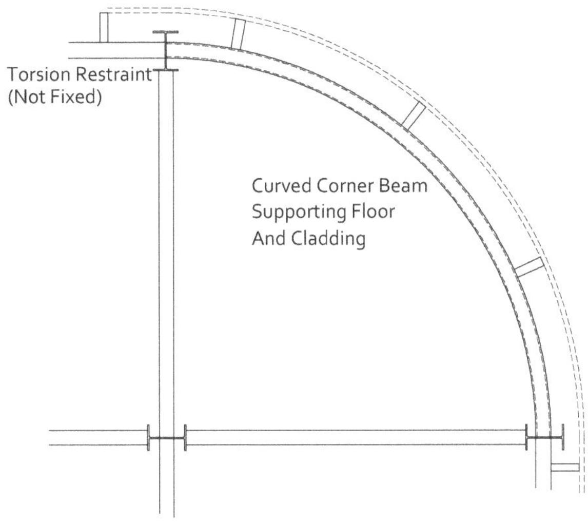

Figure 2.9.16 Curved Beam with Cladding Loads

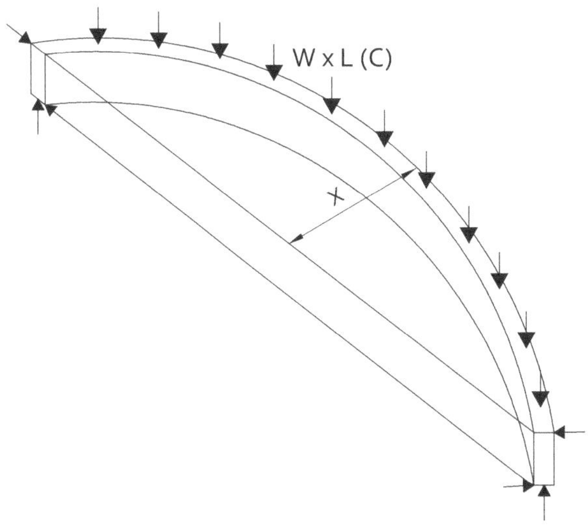

Figure 2.9.17 Beam Offset Loading

Approximate Assessment of Torsion

The following example was checked:

- **200x100x10 RHS**
- Based on a beam with radius 2m on a quarter circle. This measures approximately 3m length on the curve.
- A UDL loading of **10** kN/m
- Pinned end supports with respect to bending and restrained torsionally.

Estimating Torsion

In example illustrated by **f** the torsion may be estimated by:

1. Measure the chord offset from the beam centre line (maximum distance from the line shown in **figure 2.9.17**).
2. Work out the total loading on the beam (W x L (C)).

Find torsion. Half of this is assumed to act in calculating the torsion

(i.e. T = W x L(C)/2). L(C) is the length of the chord on the curve.

Find Torsion

Total load for **figure 2.9.17**,

Where....x = 0.60m, L(C) = 3.1m, w = 10kN/m

Torsion at ends (approximately)
= 10 kN/m x 3.1m x 0.6 /2
= **9.3kNm**

Output from a computer model, using the same parameters gave **M**= **8.3** kNm

(overestimate compared with computer model)

Floor with a Hole

In a museum in Leeds a circular floor at the top of a tower was designed to have a viewing hole in the centre. The floor is supported around the outer ring. The inner circumference was unsupported.

The floor works by means or torsional resistance and bending moment connections in the radial or spoke beams. Both circular beams are subjected to bending and torsion. See **figure 2.9.18**.

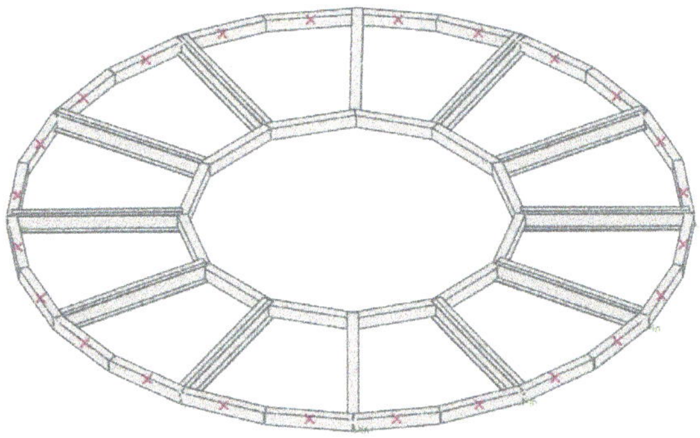

Figure 2.9.18 Gallery Floor
Royal Armouries Museum Leeds

Ring Beams

Ring beams are used to tie together outward thrusts in frameworks and structures such as domes. They resist the forces internally such that there are no outward thrusts on any external structure. Imagine a dome with a ring beam resisting the thrusts (referring to **figure 2.8.20**).

Resolution of an outward pressure (or line load) on a ring beam is shown in **figure 2.9.19**.

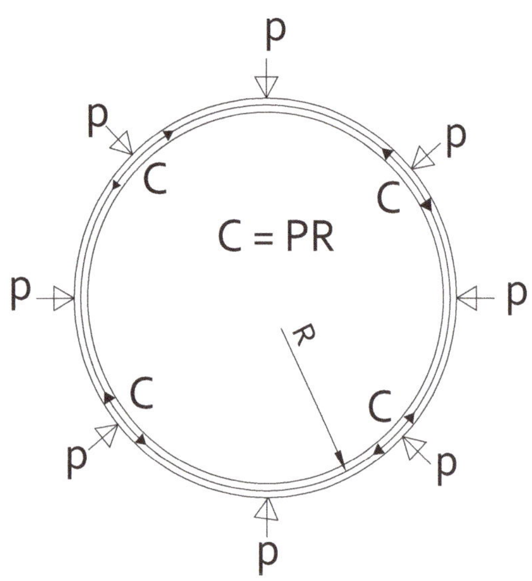

Figure 2.9.19 Ring Beam with Uniform Pressure

Compression Ring

For a ring beam with evenly applied inward pressure of **P** (expressed as force/length – kN/m):

Compression Force in Ring
C = P x R ...Internal Compression

A similar calculation (only in reverse – tension) was carried out in **chapter 2.7**. See tension roof calculations and **figure 2.7.19**.

As long as the pressure is evenly distributed ring beams are very stiff in compression or tension. There is no internal bending within the beam.

Tension Ring 45° Cut-Off (Dome)

To demonstrate how effective an iron chain or ring beam is in resisting the outward thrust of a large dome an example is taken from the 450 cut-off dome in **chapter 2.8** (**figure 2.8.20**)

See **figure 2.9.20** and subsequent calculations.

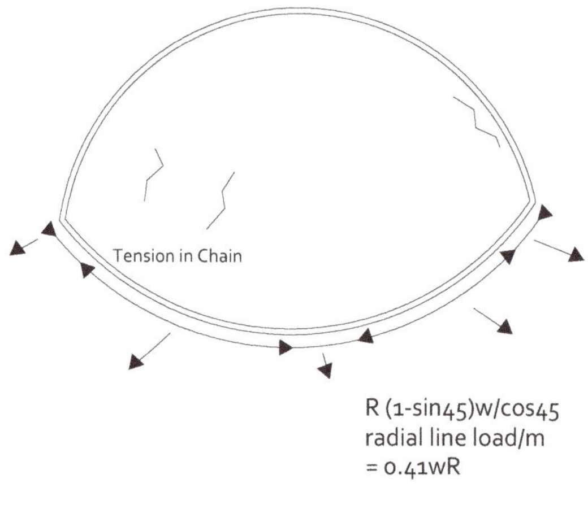

Figure 2.9.20 Dome with Tension Chain

Calculation of Chain Resistance to Thrust (Ring Beam) - uniform load

Taking the example in **figure 2.8.20**,

Loading: $w = 4kN/m^2$

(equivalent to about 150mm thick stone and some snow).

A chain is used to hold the bottom thrust:

Dome Radius, say 15m

The uniform outward thrust in the bottom of the dome for 45° cut-off is:

0.41wR = 0.41x4x15 = **25kN/m** exerted as an outward radial load.

Then the tension (reverse of the compression in **figure 2.9.19**):

T = PR = 25kN/mx15 = 375kN tension

If a tie bar has a yield strength of $210N/mm^2$ then the diameter required may be calculated by using the stress formula $P/A = 210N/mm^2$

If A (bar) = $\pi.d^2/4$ (Area) and the load,

P = 375kN then d= $\sqrt{}(4P/(\pi.210))$ in mm

Therefore......

d = $\sqrt{}$ (4 x 375kN x 10^3/3.142/210)) = **48mm**,

say **50mm** diameter bar

This seems small for a 30m span dome.

For a cut-off angle approaching 0°, the dome at the base is nearer vertical and the thrust much reduced.

Ring Beams with Concentrated Loads

In the previous example a circular ring beam was shown to be very efficient at restraining a uniform load around the edge. A ring is very good at resisting uniform radial loads.

However a thin ring (*thickness is small in relation to its radius*) suffers very large distortions under concentrated loads – a fact that is easily verified by squeezing a hola hoop

Briefly the formulae for a ring were obtained from a stress analysis book. The compression distortion of the ring selected in the previous example follows:

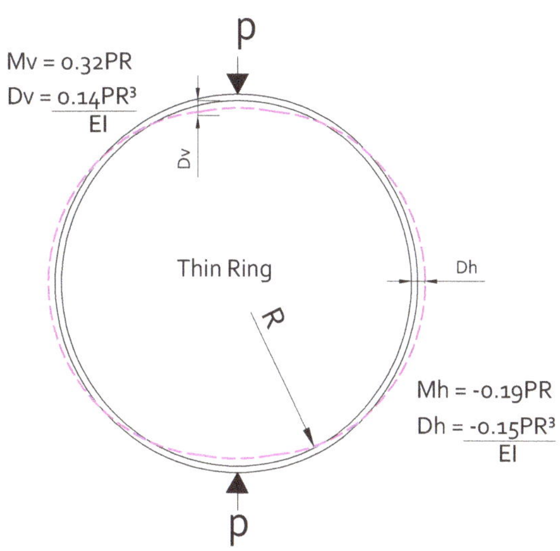

Figure 2.9.21 Ring Beam with Concentrated Load

Calculate Distortion of a 15m radius Thin Ring with point load:

Taking the example in **figure 2.9.20** and **2.9.21**, the ring, say comprises a circular hollow section (CHS) 140mm diameter x 3.2mm thick.

Area of CHS = 13.7cm^2
Moment of Inertia (I) = 320cm^4

Radius = 15m

Force applied (P) = 1kN....about the weight of one person

Using the equations in **figure 2.9.21**....

Bending Moment (V) = 0.32PR
= 0.32x1x15 = **4.8kNm**

Distortion (V) = $0.14PR^3/EI$

V= $.14\times1000\times15000^3/210\times10^3/3.2\times10^6$

= **703mm** (0.7m).....very high

The bending moment (Mh) and distortion (Dh) are similar in magnitude on the quarter segment location shown.

Rectangular Concrete Ring Beam

Figure 2.9.22 shows an engineering problem of another kind. An excavation for a basement to a new property in sandy ground is retained with piles which are concrete contiguous piles: bored close together to stop the earth and gravel being pushed between the piles.

The piles are bored by *cfa* (continuous flight augur); drilled into sandstone which grips and holds the bottom of the piles.

A concrete capping beam is cast on top of the piles – which is essentially a ring beam – a continuous cast with reinforcement throughout and linking the corners.

Earth forces comprise (generally:

1. Lateral thrust – sideways pressure from the ground. Rather like piled grain pushing against the side of a storage bunker.
2. Water Pressure. Sometimes the ground water table is at a height above the excavated level. It will exert a head of water on the side of the walls.
3. Surcharge – the weight of excavating plant and traffic pushing on the ground near the excavation will exert a pressure in the ground with a net effect pushing sideways on the face of the excavation.

The piles act as *vertical cantilevers* with the side loads acting, but are generally insufficiently fixed or stiff enough to avoid deflecting by a large degree sideways. This is not desirable and potentially unstable. A conventional means to hold the piles is to prop the top of the excavation. See **figure 2.9.23**.

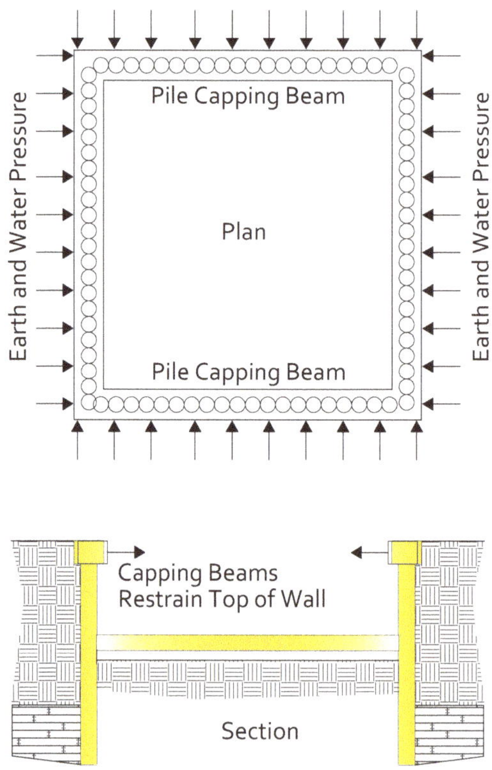

Figure 2.9.22 Retaining Wall Ring Beam

Problems with Props

Installing props imposes two major problems:

1. They need to be installed in stages. The ground cannot be excavated as soon as the capping beams and piles are constructed. Also the ground slab and foundations inside the excavation need to be excavated and cast in place. This is a tricky carefully sequenced operation.

2. Once in place they impede construction labour and materials movement and storage.

Since this particular excavation is not too large it was possible to design the capping beam to restrain the top of the excavation by designing it for the earth forces.

The bending moment diagram is shown in **figure 2.9.24**. Each side is similar to a fixed end beam (ref. **figure 2.5.1**.).

Another way of looking at the problem is to imagine cutting the ring beam and straightening out the corners; each corner being a support (partially fixed at the ends). The beam force diagram will be that for a continuous beam.

In conclusion in the ring beam solution was used in a basement project where the length of the sides was sufficiently short to utilise a reasonably sized reinforced concrete capping beam as a ring beam.

Omitting props (**figure 2.9.23**) was a great advantage in the small basement.

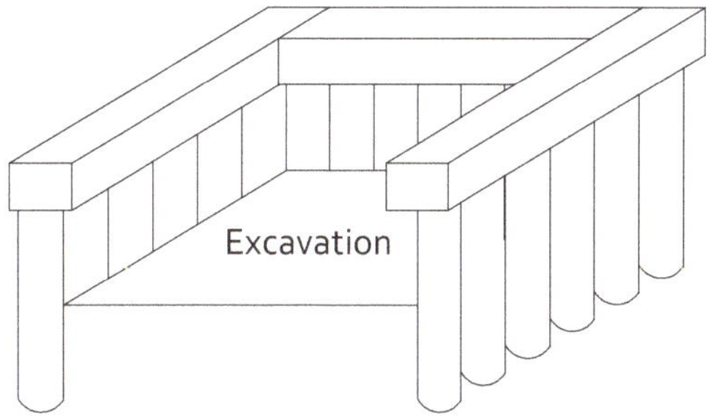

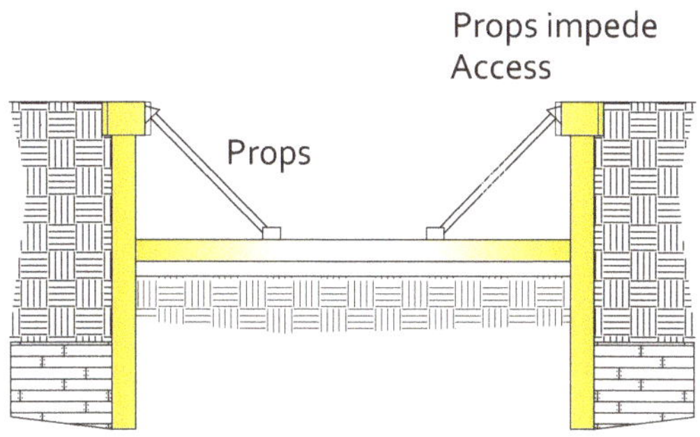

Figure 2.9.23 Traditional Propping to a Basement

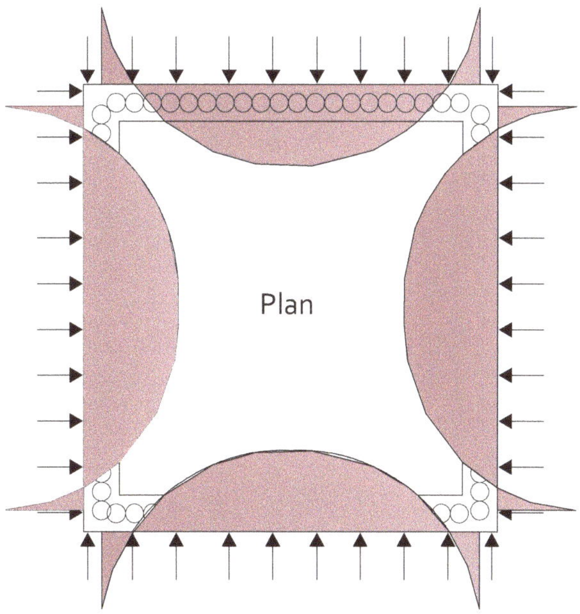

Figure 2.9.24 Bending Moment Diagram

Chapter 2.9

Quick Note Summary

1. Torsion is the product of force x lever arm (La.P).
2. The rotation of a beam under torsional load (with fully fixed ends) is TL/GJ.
3. Torsion causes shear stresses in a section (expressed as a shear flow in terms of stress analysis).
4. The shear modulus G is related to the elastic modulus E by the equation: $G=E/2(1+v)$ where v is poisson's ratio.
5. Closed Sections are much stiffer than open sections – hollow section beams are widely used for torsional loading.
6. Good practice is to *design out*`torsion; such as by adding elements which absorb the eccentric loads or bending moments.
7. Curved beams can be used with pinned ends as long as the supports are restrained from rotation in the direction of applied torsion loading.
8. Thin circular ring beams are very stiff when a uniform radial loading is applied but are very flexible and subject to high bending with loading which is concentrated or not uniform.
9. Thicker ring beams, often using reinforced concrete, are useful for resisting side thrusts or prevention of excessive spread.

Chapter 2.10

2.10 Plate Structures

It was not until rolled sheet steel could be produced to a consistent quality in the 19th century that thin steel plates could be put to widespread structural use. The widespread use of reinforced concrete with sophisticated design methods did not really arrive until the 20th century. Sophisticated design of thin-plated structures is common to the shipbuilding, process and aerospace industries but tends to be a specialist area in construction. This chapter pulls together a range of applications under the umbrella of plate structures: two-way spanning plates, reinforced concrete slabs, masonry walls and stressed skin timber and SIPs. Brief analysis is reviewed to illustrate the benefits of plate action.

Benefits of Plate Structures

Plates may be employed as economic structures. They are robust and may be broken down into modular prefabricated units connected together on-site in fast-track methods of erection. In common plates may be defined as having the following benefits:

- Two-way spanning action (bi-directional bending) floors and walls: steel or concrete.
- Diaphragm action: the ability to hold together structural components in horizontal or vertical directions in shear: timber floor decking.
- As vertical elements act as deep in a vertical orientation. See **chapter 2.11**.
- To act as core elements in the lift shafts and stairwells of buildings – often providing the stabilising structure for tall buildings. See **chapter 2.11** and **chapter 2.12**.

Properties of Plates

For simplicity, plates may be defined as having the following common properties:

- They are shallow in proportion to their span – a ratio of thickness to span is usually in the region of 20 or more when loaded perpendicular to the thin side (i.e. not like a deep beam).
- They are isotropic: the material has the same properties in all directions.

The material is elastic – non-linear, and plastic behaviour is necessary to design reinforced concrete structures, but for this purpose, elasticity is assumed.

Two Way Spanning Action

A simple example of plates is shown below.

The equations can become very complex, depending on the support conditions.

This example will demonstrate the *two-way spanning* the potential of plates.

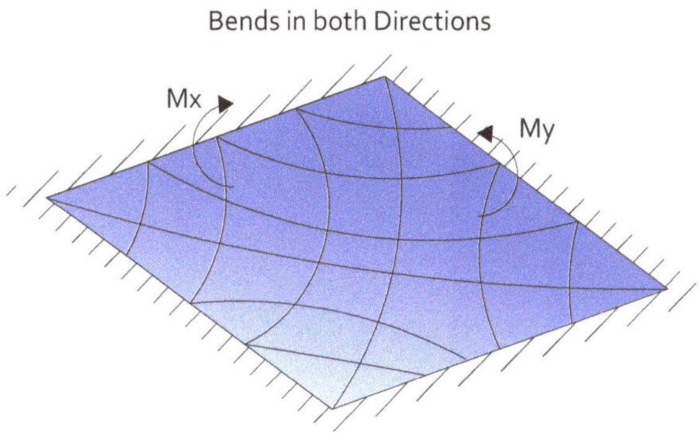

Figure 2.10.1 Plate Two Way Spanning Action

Analysis of a Square Plate with UDL

A square plate with a downward uniform load per unit area spread across the surface was compared for two different support conditions:

1. Supported on four sides with a pinned support (*rotation allowed and hence no support moments*).
2. Supported on two edges only (pinned).

The bending moment and deflection equations are displayed for each case in **figure 2.10.2**.

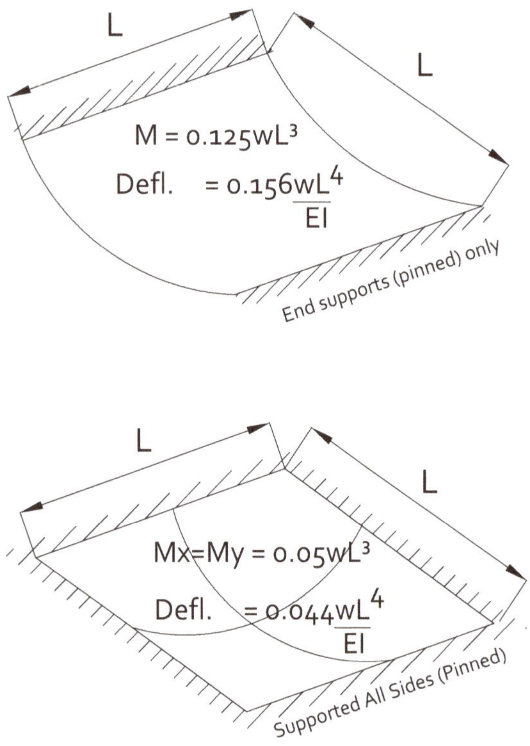

Figure 2.10.2 Square Plate Formulae

By supporting on four sides, the plate has a *two-way spanning action*. In other words, it may bend in two directions since the properties of the material are identical in two axes.

By adding two supporting sides:

- The bending moment is 40% less than in the case of spanning in one direction.
- The deflection is 28% less than for the one-way span.

The improvement in stiffness and reduction in flexural stresses is better than simply halving the deflection and moment.

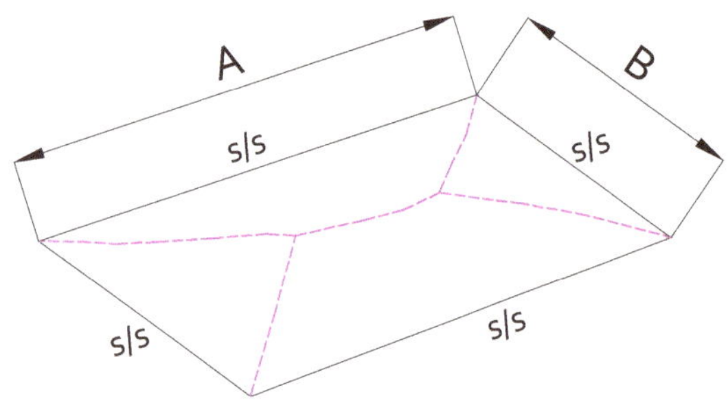

Figure 2.10.3 Yield Line Pattern

Concrete Slabs

Concrete slabs pose a problem with elastic analysis. In the ultimate limit state, the reinforcement is designed to remain ductile at failure, and the concrete is allowed to form cracks in the bottom tension zone of simply supported or continuous spanning slabs.

Cracked concrete has approximately half the modulus of uncracked concrete. The material tends to behave plastically in the design condition.

This is not too much to fear for the designer since there are many design aids and charts to assist in analysing the properties and behaviour of concrete structures.

A Danish engineer, Yohansen, developed the theory of yield line behaviour method in the 1960s which analyses the slab by energy methods. In short a given slab is assumed to form cracks in which the analyst plots a pattern assuming crack failure in the slab. A typical yield line for simply supported edges is shown in **figure 2.10.3**. The bending moments in the slab are then calculated from the amount of energy needed to form the crack.

It is a very good method for analysing slabs without too much calculation. The designer has to select a mode of failure, which is always a good thing as a visual method of analysis. However, with the advent of computing and software, finite element and plate element methods of analysis are available for slabs, enabling the most complex of supporting arrangements and slab geometry to be analysed.

An engineer should not be deterred from making approximate checks when using design charts or finite element analysis.

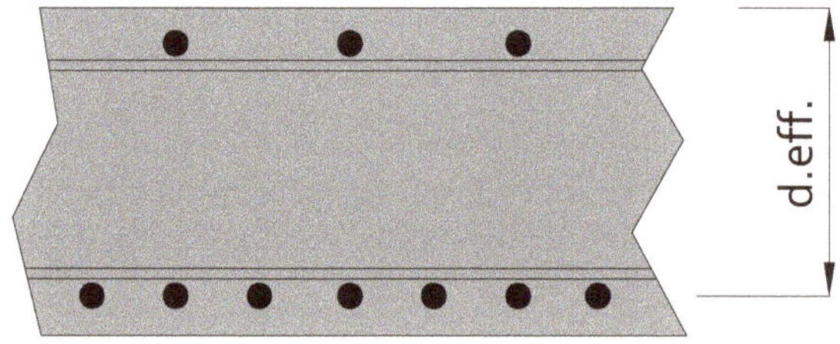

Figure 2.10.4 Concrete Slab Effective Depth

Elastic Analysis Assumed Properties

Concrete slabs and beams may be checked by hand calculation assuming the following parameters:

1. The depth or thickness is the dimension to the centre of the active bar in tension (bottom bar for midspan support and top bar for support – if fixed or continuous). See **figure 2.10.4** – a slab is similar.

2. The cracked modulus of the concrete is used for long-term loading – assume half the uncracked modulus (generally around **12-14x10³** MPa or N/mm²).

3. For short term loading use **24-30,000** MPa (N/mm²) for conditions such as dynamic loading.

For slabs, plate formulae (for instance, **figure 2.10.1**) may be used for an approximate assessment, but tables are generally available in national codes.

For supporting beams, the normal equations for simple beams described in **chapters 2.2** to **2.4** are applicable to approximate analysis.

This method is very useful for checking the output from a plate or finite element analysis.

Masonry Walls

The analysis of masonry walls, with side loading from wind or other lateral loads, is similar to that of concrete slabs.

Masonry in modern usage may be reinforced or unreinforced. In general use, it is not reinforced; the strength relying on the bonded mortar joint between bricks or blocks. Masonry exhibits *anisotropic* behaviour; a term that means it has different properties in the vertical and horizontal directions.

Masonry codes of practice contain tables based on the aspect ratios (ratio of length to height of wall) to obtain empirical factors for analysis. Software is also available.

This section compares a simplified form of checking masonry in which it is assumed to arch. The principles of **chapter 2.8** are used, and a wall is assumed to span in one direction. This is to be compared with an analysis using empirical formulae from charts based on the wall aspect ratio.

Arch Action of Masonry Walls

For estimation purposes a wall can be checked for one way spanning action assuming that the mortar has no tensile strength. The problem then becomes one of *arch* action. Loading is by wind or other lateral action.

A thrust line is traced in the masonry within the middle third and the internal compression at failure equated to the compression strength of the masonry unit with mortar (published in codes of practice or tables).

Estimate of Masonry Wall Strength

Suppose the wall in **figure 2.10.6** had the following parameters:

L (distance between side walls)	= 3m
W (wind loading)	= 1kN/m²
Thickness of wall	= 140mm
Strength of masonry	= 1.4N/mm²

Arch Formulae for One Way Span Action

Using the arch formulae in **figure 2.8.7** (lever arm in section 2/3 thickness):

Rh (per metre height) = wL²/8h
= 1x3²/8(2x.14/3) = 12kN/m

Rv (per m height) = wL/2
= 1x3/2= 1.5kN/m

Thrust (compression in brick)
= √(12²+1.5²) = 12kN/m

....Rv may be ignored (the angle of arching is acute).

Assume compression on 1/3 face of brick

Compression stress on brickwork
= 3xthrust force (area of brick)
= 3x12x10³/140/1000
= **0.26 MPa (N/mm²)**

...this is less than the masonry strength...the masonry has sufficient strength.

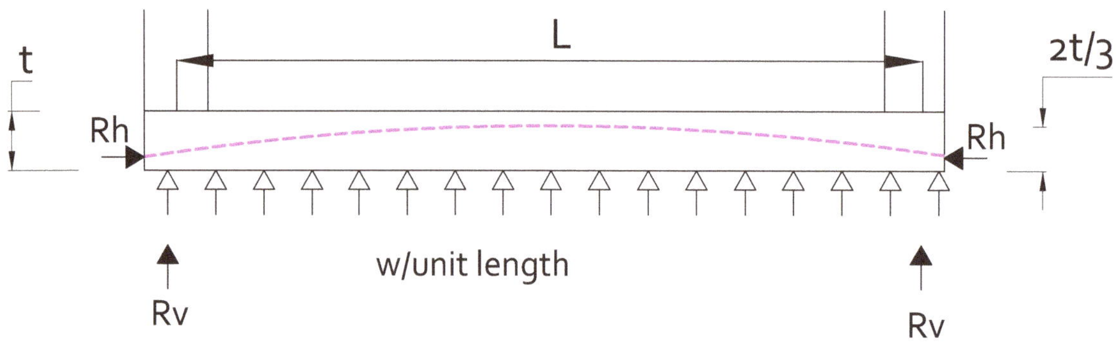

Figure 2.10.5 Masonry Arch Action One Way Span

Wall Two Way Span Analysis from Limit State Tables

Walls, in practice, are supported along the bottom (see yield line in **figure 2.10.3**).

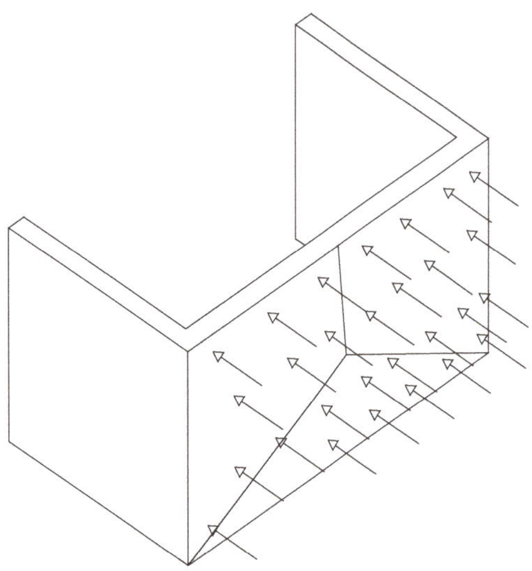

Figure 2.10.6 Masonry Wall Yield Line Two Way Span Action

The base of the wall provides an additional supporting edge, which effectively strengthens the wall over the two edge support conditions at the *return walls* shown in **figure 2.10.6.**

Find Maximum Stress in Bending

Referring to tables in the UK Standard for masonry show, for a 3m long wall (3m high) with side supported on the bottom and side edges, the bending moment is quoted as:

M (middle) = $\mathbf{0.045wL^2}$

Using the parameters in the previous example, the bending moment is:

$M = 0.045 \times 1 \times 3^2 = 0.40\text{kNm}$

The stress in the wall is found in the formula M/Z, where

$Z = bd^2/6$ (Modulus rectangular Section)

Therefore maximum compressive stress, is:

$\sigma c = 0.40 \times 10^6 \times 6/(1000 \times 140^2)$
= **0.12 MPa (N/mm²)**

Comparing this with the value of **0.26MPa**, the arch action method (previous example) gives twice the stress of the more refined (2 way span) analysis.

In other words, the arch method is a safe first approximation without the use of tables or more complex methods.

Sandwich Structures

Modern structural materials take advantage of the properties of materials in composite forms. A typical type now in widespread use is the SIP panel (or *structurally insulated panel*).

A typical SIP panel section is shown in **figure 2.10.7**.

It comprises two components: outer sheets of *orientated strand board* (**osb**) and an inner core of foam which is commonly *polyurethane foam* (pur) which has a high bond strength to the osb skins.

The panels with pur foam are manufactured by pumping foam between two sheets of osb, which then expands and sets, bonding with the inner faces of the osb. Other core materials such as polystyrene are used and sometimes thin gauge steel sheets for the outer faces.

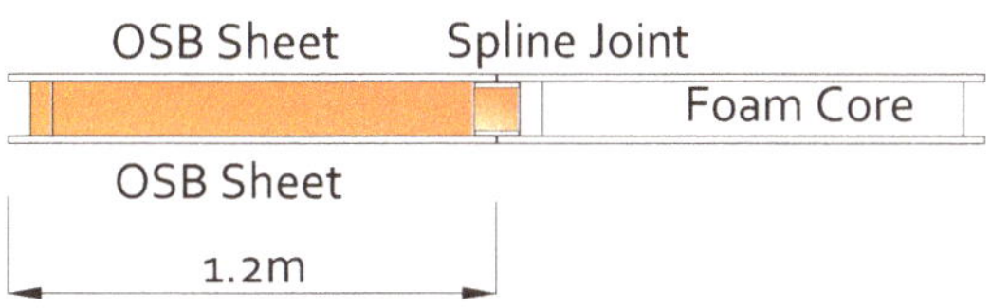

Figure 2.10.7 SIP Panel Section

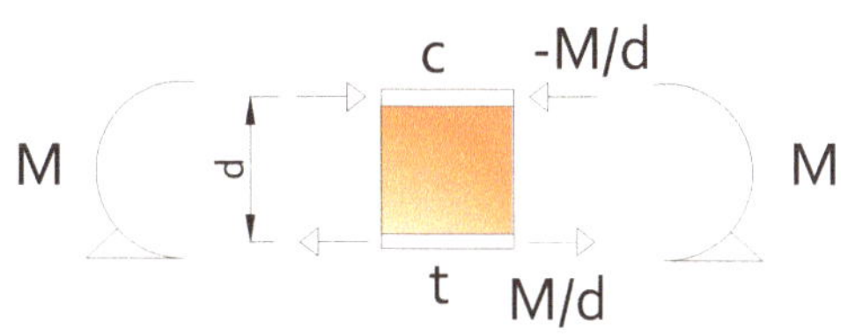

Figure 2.10.8 Bending Stressed Skin

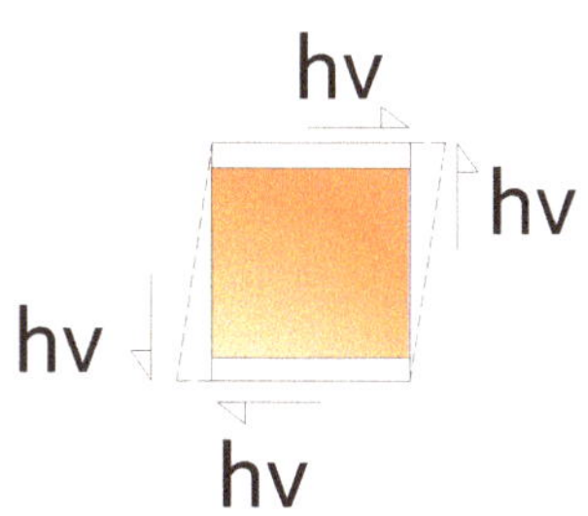

Figure 2.10.9 Stressed Skin Shear

Stressed Skin Action

A SIP panel acts as a *stressed skin* - the osb sheets as tension and compression elements - like the top and bottom chords in a truss. The faces experience maximum tension and compression (as in a simply supported beam) in the centre of the span of a panel.

See **figure 2.10.8**. – – where d is the depth (*from face to face*).

At the supports, where the shear is at a maximum for a uniformly distributed load, the shear flow in the section is at its maximum in the core. See **figure 2.10.9**. The foam tends to form a lozenge shape or shear distortion.

SIPs Panel Construction

On a roof or floor, the SIP panels are fixed side by side with connecting *splines* (see **figure 2.10.7**.).

Figure 2.10.10 shows a house internal view with SIP walls and roof panels.

A particular benefit of SIP panels for house construction is their plate action. The roof panels provide two-way spanning action and a diaphragm to stiffen the building.

A note of caution is that the span action perpendicular to the panel length does not have as great a stiffness or bending strength. Shear distortion of the foam core causes long-term creep in the panels, which needs to be checked. The walls act as shear walls, which, like the sides of a cardboard box, give rigidity or stability to the building.

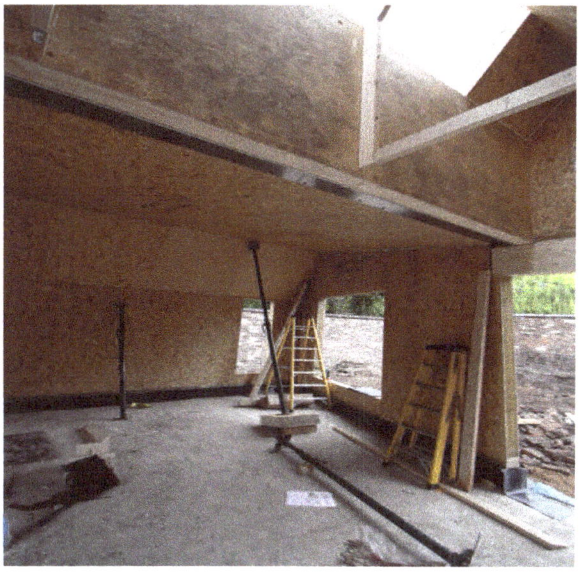

Figure 2.10.10 House Constructed From SIPs

Chapter 2.10
Quick Note Summary

1. Plates exhibit two-way spanning action.

2. A square plate with 4 supports will have less than 50% of the bending moments and deflections of the same plate with 2 edge supports. This is because of a stiffening effect of the plate geometry.

3. Concrete slabs should be analysed as cracked sections when analysed as plates.

4. Masonry walls are *anisotropic*: they have different properties in different directions.

5. Masonry walls may be estimated for strength by arch action.

6. For two-way action in masonry walls tables are usually provided for different support conditions to undertake empirical checks.

7. Modern sandwich panels such as SIPs joined together behave with two-way spanning action; although the *across* joint spanning action is not necessarily as strong or as stiff.

Chapter 2.11

2.11 Deep Beams and Diaphragms

Deep beams are, in essence, plates turned on their sides; a crude analogy but a visual aide in picturing the action of deep beams. Sometimes, walls on the upper floors of buildings or over basements may be designed as dual action deep beams. This chapter reviews a concrete deep beam, a SIP wall spanning over a void and floor diaphragms, all examples of deep beam action.

Concrete Deep Beams

Figure 2.11.1 shows a concrete deep beam. The internal load path or stresses in the section may be validated by a method called the strut and tie method – which is essentially a truss analogy. It is a good method to use, or at least to practise, since it forces the engineer to consider carefully the balanced load path within the structure.

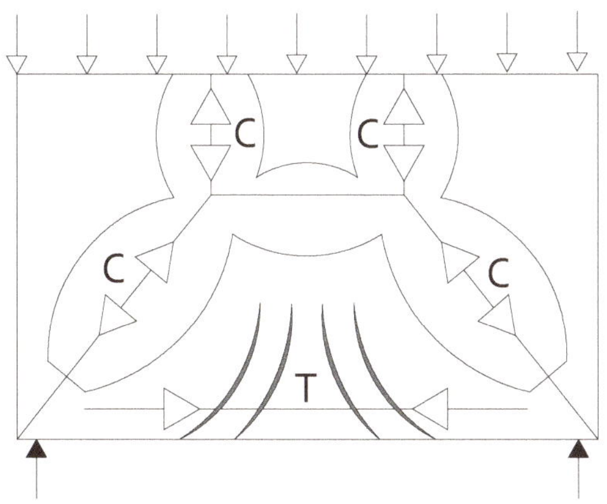

Figure 2.11.1 Concrete Deep Beam

The uniform loading on the beam is carried by the deep beam by a diagonal compression zone in the concrete. At the bottom of the beam, the concrete is in tension when subjected to the diagonal compressions in the beam.

Reinforcement bars are placed in the bottom with sufficient sectional area to resist the calculated tension without snapping the bars. As long as the concrete with the selected area is not stressed beyond its compression strength and sufficient bars are placed in the bottom zone, the section will be strong enough.

Estimate of Reinforcement Quantities

A very simple method to calculate the number of bars in the bottom and the number of vertical bars in the concrete for an initial estimate is to work out the effective lever arm. Deep beam theory has shown that the effective depth (or lever arm) for a deep beam is the product **0.62xdepth**. See **figure 2.11.2**.

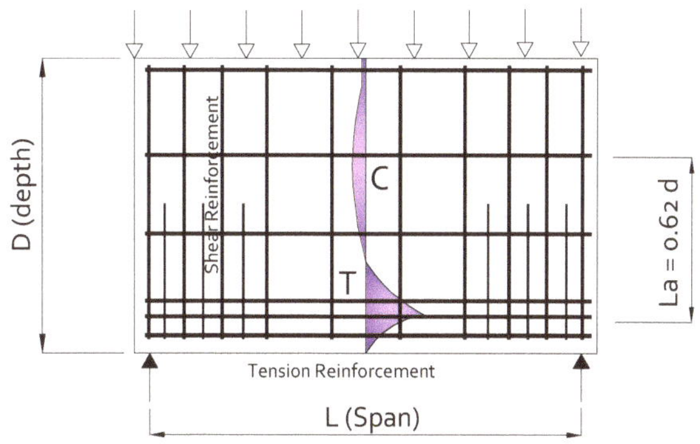

Figure 2.11.2 Deep Beam Reinforcement

The section will need enough vertical reinforcement bars to resist the shear of the section by this method. A shear stress calculation is carried out over the effective depth.

Bending Moment Calculation	
Span, L of beam	= 4m
Loading on beam - factored load	= 200kN/m
Width of beam	= 400mm
Depth of beam, D	= 2m
Effective depth= 0.62 x 2.0 = **1.24m** (D_{eff})	
Bending Moment = $wL^2/8 = 200 \times 4^2/8$	= **400kNm**

Calculate Bar Area

Tension in bars = M/D_{eff}
= 400/1.24 = 324kN

Bar strength (tension) = $500N/mm^2$

Therefore, bar area needed
= $324 \times 10^3/500$ = **$648mm^2$**

A 16mm diameter reinforcement bar has an area of $201mm^2$, so to provide this area of bar:

Number of **16mm** bars needed
= 648/201 = **3.2** ....(round up to **4** bars)

Shear Reinforcement Estimate

Shear force at support
= 200x4/2 = 400kN

Shear stress at support (V/b.deff)
= $400 \times 10^3/1240/400$ = **$0.80N/mm^2$**

The allowable shear stress in a concrete section is approximately $0.40\ N/mm^2$.

So vertical reinforcement bars will be needed.

This is a simplified calculation. Detailed calculation is generally carried out with finite element design on a computer or strut and tie type analysis, which is more rigorous. It is nevertheless a good method of approximation to check the results.

Aside

Anchorage of Bars

To ensure correct anchorage of bottom reinforcement, the tension in the bottom bars is more or less assumed to be a constant. The bars are bent up at the ends to achieve anchorage. This is the distance beyond the support where the bar needs to be anchored into the concrete to ensure it would not pull out of the anchorage point under full load.

As a rule of thumb, the anchorage length is 35-45 x diameter of the bar for a straight bar.

In other words, a 20mm bar will need to be bent around at the support and terminated 900mm past the anchorage point as a worst-case assumption.

Anchorage may be achieved by hooking the bar around in a U-bend or into an L shape, which will need less than the 900mm.

A SIP or Sheathed Deep Beam

Another form of deep beam is achieved with SIP walls or timber frame stud panels with sheathing. In contrast to the previous concrete beam example, checks are usually focused on the shear flow or interaction between the skin and internal timber studs.

A full-height SIP wall made up of three panels spliced together is shown in **figure 2.11.3**. A shear flow pattern is shown around the edges. In the centre of the beam, the shear is nominally zero for a uniformly distributed loading, increasing to a maximum at the supports.

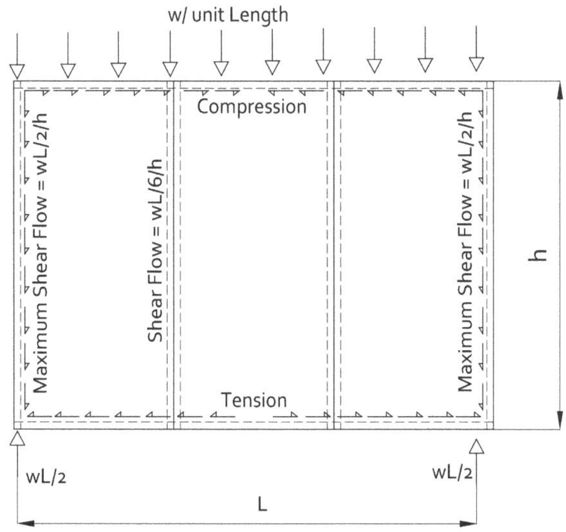

Figure 2.11.3 SIP/Stud Wall as Deep Beam

Note that the arrows are parting at the bottom of the panel (tension) and pushing together at the top (compression); the timber plates acting as chords – as trusses referred to in **chapter 2.6**. Or viewed another way, they are analogous to flanges in a beam.

A SIP panel is designed to have continuous timber plates or joists nailed and sometimes glued into the recesses in the panels.

SIP or Sheathed Beam Shear Flow

The shear flow is a maximum where the shear force is a maximum: at the support. To ensure that the panel has sufficient strength, shear, which is carried within orientated strand board (osb) skins, needs to be transferred to the end timbers (vertical timbers built into the end recesses). In design terms, this means that the engineer must ensure there are sufficient nails to fix the skins to the edge timbers.

The SIP panels have a permitted racking strength (or shear strength), depending on the manufacturer. Standard size panels are tested by putting a lateral load onto a panel which is fixed to the floor at the bottom. This gives a maximum design in-plane shear loading for the panel.

SIPs and Timber Frame Typical Racking Strength

As a rule of thumb SIPs quoted racking strengths are in the order of **3-6**kN/m, timber frame sheathed panels **1.7**kN/m (*sheathed on one side only*) and **3.4**kN/m (*sheathed both sides*).

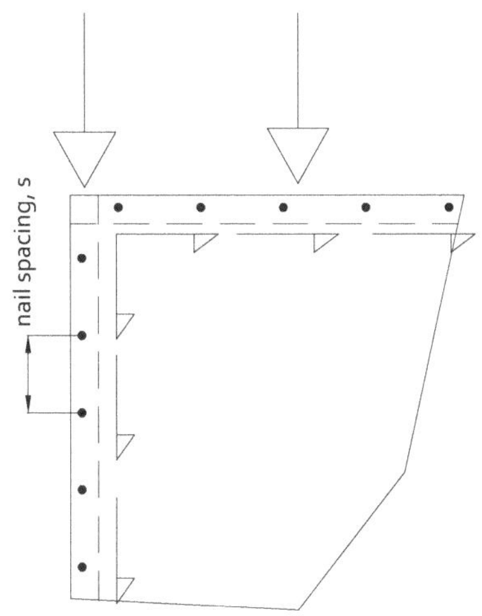

Figure 2.11.4 Local Detail at Corner

SIP or Sheathed Deep Beam Calculation

If a SIP panel is designed as a deep beam, the maximum in-plane shear loading should never exceed the manufacturer's quoted shear or racking strength. In the example below, the maximum racking load quoted by the manufacturer is 6kN/m.

Calculation For a SIP Deep Beam

Panel Height = 2.4m
Beam width (3 panels wide) = 3.6m
Loading = 8kN/m
-including the weight of the wall

Racking Strength = **6kN/m** manufacturer (*BBA Certificate*)

Bending Moment on wall = $\mathbf{wL^2/8}$
= $8 \times 3.6^2/8$ = **13kNm**

Support shear load = **wL/2**
= 8x3.6/2 = **14.4kN**

Check Shear flow (racking) at support
= 14.4/2.4 = **6kN/m**

The panel is at its maximum permissible shear.

Timber Frame Walls

A similar principle may be applied to timber frame walls, which are sheathed with a shear-resisting skin such as plywood or OSB.

Nail spacing and sizes need to be checked for the shear flow.

Floor and Roof Diaphragms

Figure 2.11.5 represents a plywood floor (or osb) on timber joists. The left-hand wall at the first floor in a house is subjected to a wind force from the sides of the house. As a cross wall it sits only on posts at ground floor. Because the posts are not rigid in terms of lateral loading the wind load takes the stiffest load path down to ground. In this case the plywood floor (with staggered sheets to improve the diaphragm strength) carries the wind load P across the house.

In other words, like the previous example, the floor is a deep beam on its side. It transfers the wind force **P** by means of its nailed joints into the joists. At the other end of the floor is a ground floor cross wall. This cross wall collects the load **P** and transfers it into the foundations (like a deep beam).

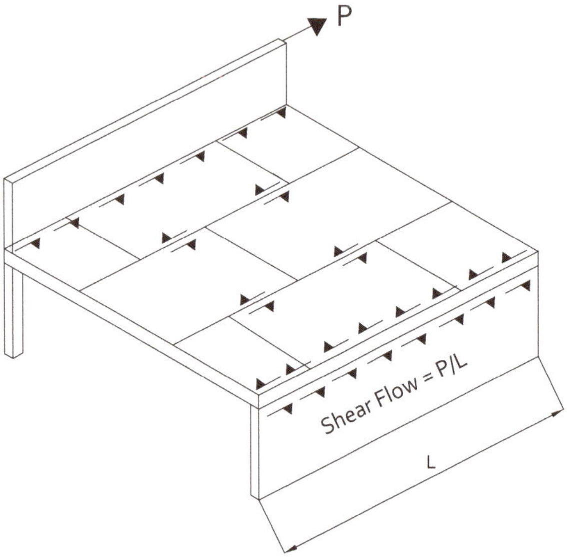

Figure 2.11.5 Diaphragm Floor

Shear Flow Calculation

The fixing of the right-hand wall, which carries the loading to the ground into the floor, should be checke.

Shear flow into wall (length L), Sv
Sv = P/L (in kN/m)

This may be used to work out the number of nails by the method in the previous example. Usually, this should be no more than around 150mm spacing, but may be less.

Aside

Building Stability

Floors are integral to building stability in the sense that they transfer shear into stabilising structures as diaphragms or deep beams. Examples of stabilising structures include:

- ***Concrete cores (lift shafts and stairwells). See figure 2.5.15 and 2.5.16.***
- ***Steel braced Frames (see frame section figure 2.5.6).***
- ***Rigid Frames. See figure 2.5.9.***
- ***SIP or sheathed timber frame walls.***
- ***Masonry walls***

Pitched Roofs as Deep Beams

Plate action in roofs to ameliorate the effects of roof spread is another aspect of plate action where roof panels act as sloping deep beams; as roof slopes increase in angle from the horizontal, the deep beam action becomes more influential under vertical loads.

Two features of roof plates are discussed briefly: nailed plywood or nailed boards onto a traditional rafter timber roof (*sarking*) and roofs formed as SIP panels laid on the slope.

Sarking

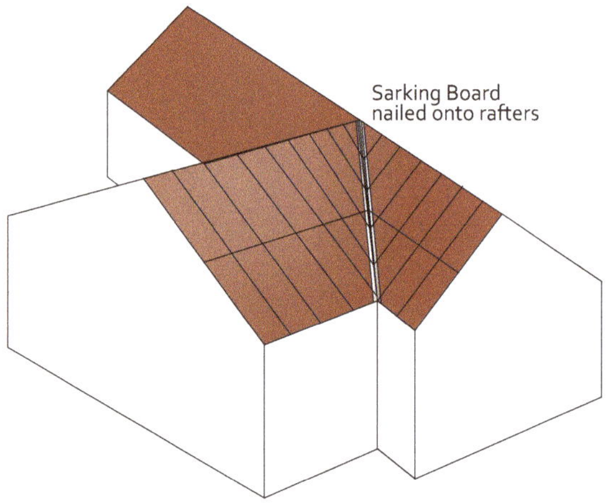

Figure 2.11.6 Sarking

Traditionally, in Scotland, for example, a sarking board is nailed to the top of joists to form a top plate across the whole area of the roof. The plywood or boarding acts in the sloping plane of the roof to resist roof spread in the manner of a vertical panel described in **figure 2.11.3**.

SIP Roof

SIP roofs act as plates, which is particularly useful on pitched roofs. See **figure 2.11.7**.

SIP panels are generally laid along the slope of the roof and jointed, as noted in **figure 2.10.7.** This creates a plate action which acts within the plane of the roof. With two OSB skins SIP panels are even more rigid in plane than roofs with sarked roofs, which are described previously.

SIP Roof Analysis

The behaviour of SIP roofs is difficult to model by hand. Using sandwich panel properties converted to idealised isotropic plates the deflections and bending in pitched roofs may be estimated prior to modelling by hand. As an example, a **150mm** deep SIP panel has an idealised modulus, assuming a homogeneous material, approximating to **120MPa** (N/mm^2) for long-term (creep) deflection. If it is reinforced in the spline joints with solid timber, this increases. **Figure 2.11.8** shows a pitched roof analysis output for a complex roof for long-term creep deflections using idealised plates.

Notice at the closest edge the maximum eaves deformation is circa 4mm. Minimum restraint was assumed by the action of the wall, demonstrating the *spread resistance* of this form of roof with *plate action*.

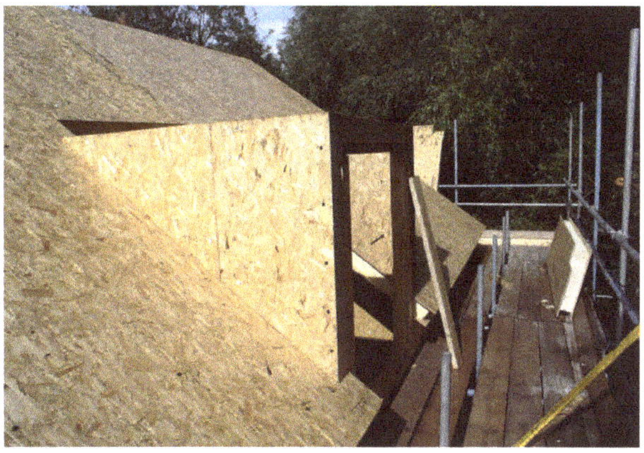

Figure 2.11.7 SIP Roof

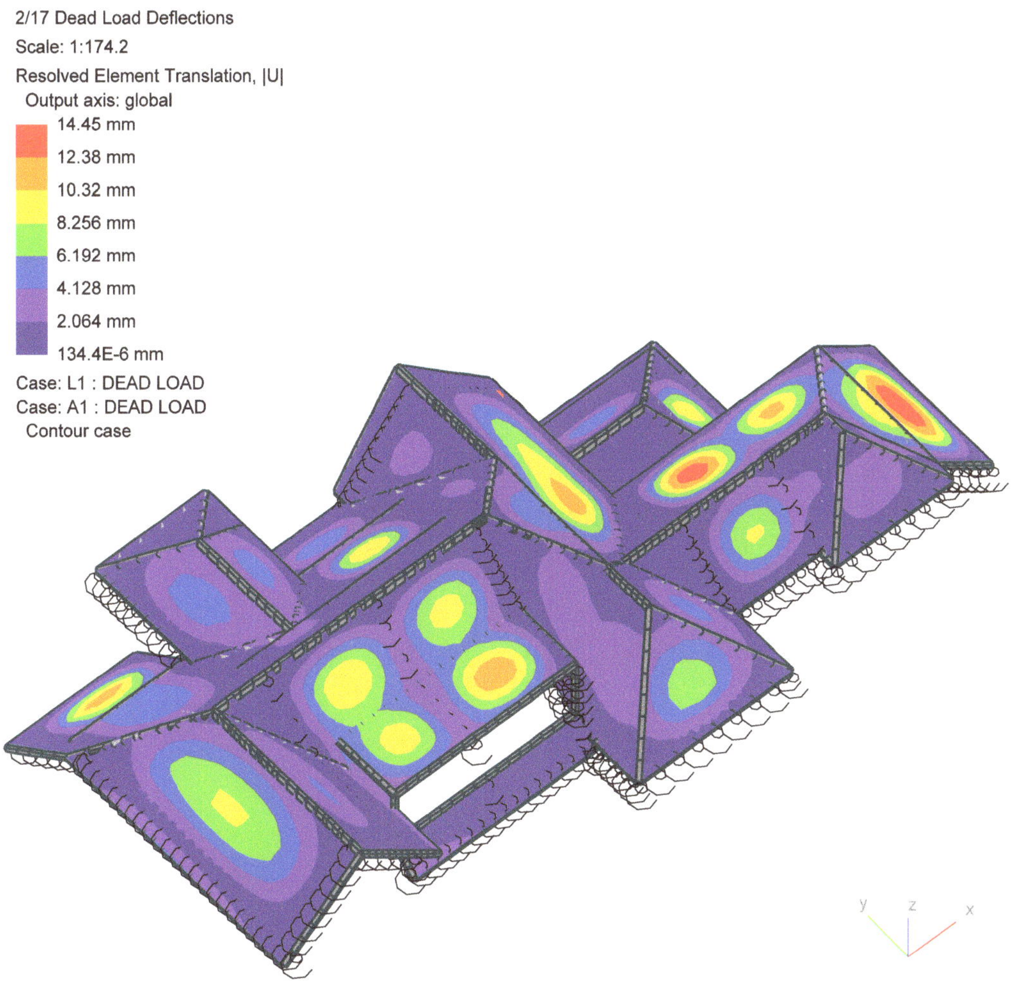

Figure 2.11.8 SIP Roof Plate Model

Chapter 2.11
Quick Note Summary

1. Deep beams in concrete are effectively analysed as idealised trusses with concrete as compression struts and reinforcement bars as tension ties.

2. Concrete deep beams need special attention to the effective depth.

3. SIP and sheathed timber frame walls may be used as deep beams. Careful checking is needed for the shear connection.

4. The racking strength of a SIP/Sheathed timber stud wall is effectively the maximum in-plane shear it can carry.

5. Floors in buildings act as diaphragms carrying the wind loading and other lateral loading and transferring it onto the stiff walls and frames of the building.

6. Sarked or SIP roofs behave as sloping deep beams in the plane of pitched roofs. They assist with spread resistance in a more robust manner than traditional rafters with batten roofs.

Chapter 2.12

2.12 Dynamics

Vibration is a subject in the textbooks and on any structural engineering course, which is too often off-putting to all but the most mathematically able. It is argued that engineers should have at least a rudimentary working knowledge of the principles, particularly with a modern tendency for clients and architects to demand long-span lightweight structures - which are especially prone to dynamic excitation. The first part recaps on

the main elements of dynamics, followed by simple rules for finding fundamental natural frequencies. A review of simplistic methods to assess peak structural response, together with examples which demonstrate how to identify areas of concern, is followed by a

short summary of some response criteria. In conclusion, aspects of building sway (dynamic response to wind loading and earthquakes) are discussed.

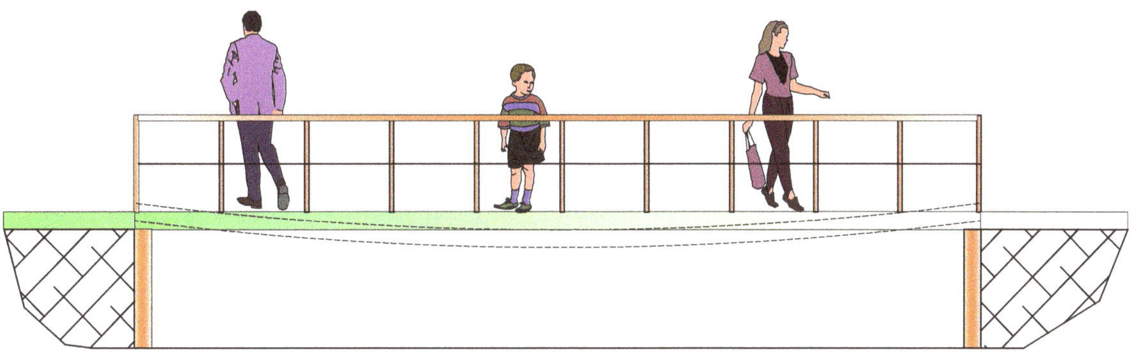

Figure 2.12.1 Dynamic Footbridge Response

Dynamic Response

Vibration is a very important aspect of understanding and designing structures. At one end of the scale, collapse and failure of buildings under the seismic vibration caused by earthquakes can be fatal. At the other end of the scale, the experience of vibrating structures may, at best, irritate and, at worst, cause anxiety to the users of buildings and bridges.

Large machinery in buildings and factories, by the nature of their power and rotary motion, can damage structures or distract the building occupants. Isolation measures may be necessar.

Tall buildings sway in the wind. This sets up a periodic, pendulum-like motion to which humans are very sensitive. The effects are akin to the feelings of seasickness in the worst cases.

Dynamics and its effects tend to be regarded as a specialist subject. Dr Aleksander Pavic and Dr Paul Reynolds run a course at Exeter University in the UK, which has a faculty with particular expertise in floor vibration and response. This kind of expertise is consulted when determining the response of building floors and frames to vibration imparted by building occupants and plants. In hospital operating theatres, some of the equipment is vibration-sensitive to even the most insignificant footfall of a single person.

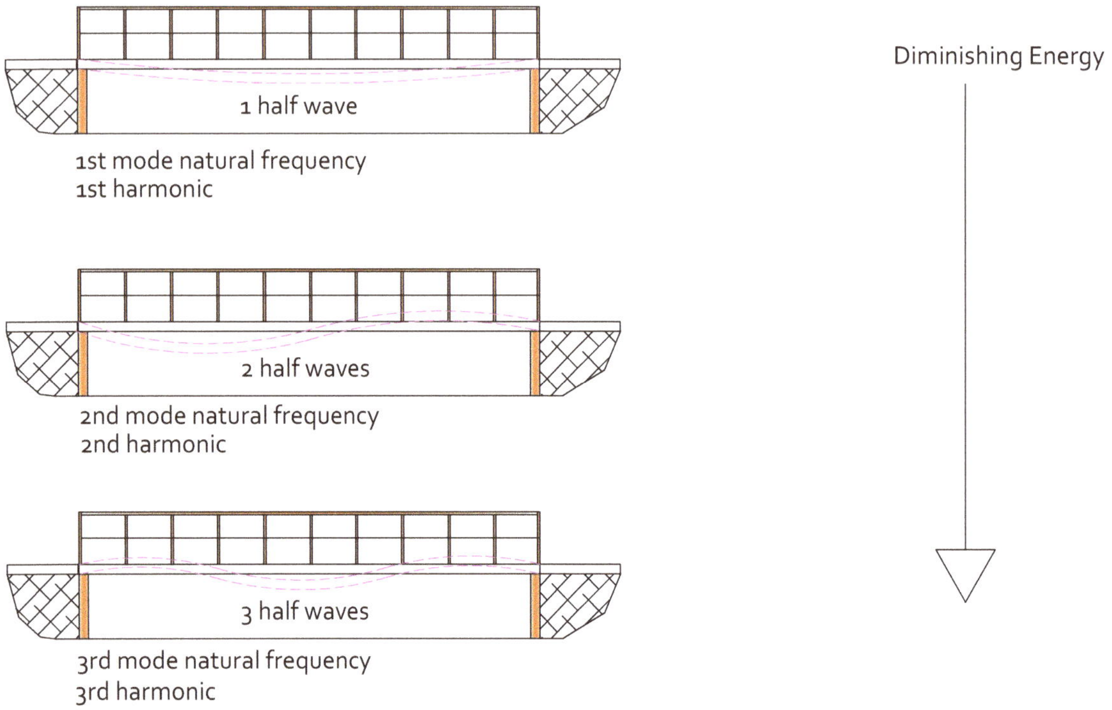

Figure 2.12.2 First to Third Modes Natural Frequency

What a General Practitioner Should Know

A general practitioner, in other words, a structural engineer working on structures such as buildings or bridges, should have a basic understanding of vibration. It is unfortunate that engineers tend to fear this aspect of design by the complexity of the subject. Engineers are practical people who want to gain an intuitive understanding of the structure at hand.

The saying: '*it is important to know what you don't know*' is the apposite. If the engineer has to design a laboratory with sensitive equipment, within these five basic steps, he or she will at least be able to undertake a scheme design in which the dynamic performance of that structure is acceptable. Having gone through these steps, at the least, meaningful discussions are possible with a specialist floor vibration analyst or engineer

A summary of the level of understanding of all structural engineers is suggested below. For serviceability analysis a structural engineer should be able to:

1. Identify and quantify the source of the vibration (footfall, say), the path (*mass, stiffness* and *vibration)* and the receiver (*end-user*). This chapter is relevant to serviceability analysis: the problems of perception rather than strength.

2. Calculate the 1st mode natural frequency of a structure: in the vertical direction for floors and lateral direction for frames.

3. Select damping for a particular structure. Damping is the degree of absorption of vibration energy in a structure – by means of walls, suspended ceilings and services, balustrading and internal (inherent) energy absorption.

4. Assess the distribution of mass (modal mass), which is most significant to the motion of the structure.

5. Select appropriate dynamic loading. Calculate a degree of response to the dynamic loading (*usually expressed as acceleration or velocity*).

The Basics of Dynamic Motion

Natural Frequency

The first step in assessing a structure is to calculate the natural frequency. Every structure has a number of natural frequencies in harmonic modes in which it will vibrate according to the waveform. **Figure 2.12.2** shows the first three harmonic modes of frequency in a simple footbridge.

Application of Static Load

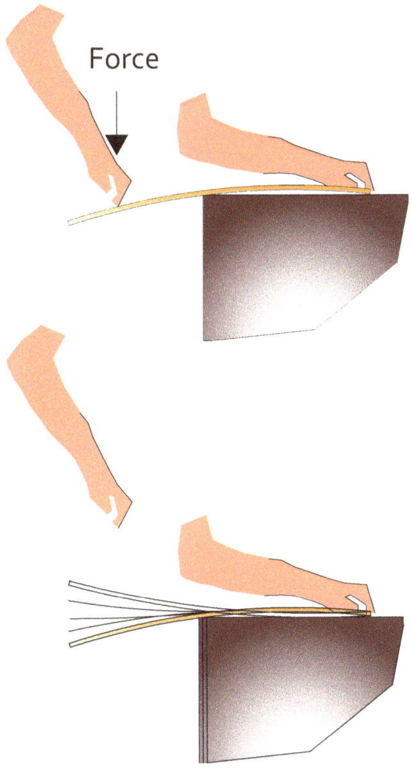

Figure 2.12.3 Dynamic Response Upon Release of Static Load

When a static load is applied to a structure, it will impart deflection. Upon release of the load, the structure releases the potential energy by vibrating, resonating in all the modes of natural frequencies with diminishing response from the first mode going up the scale.

Figure 2.12.3 illustrates a flexible ruler being made to resonate under a static load or force. All modes in which significant response is experienced should be investigated.

For the purposes of simplicity, the first mode, natural frequency will be considered in this chapter.

If the free end of the ruler is shortened, the cantilever is reduced, causing the structure to become stiffer. This raises the natural frequency, hence the pitch of the sound.

Application of Dynamic Load

This is when a periodic repeated load is applied for instance, a person walking across a bridge or bouncing on a springboard (see **figure 2.12.4**).

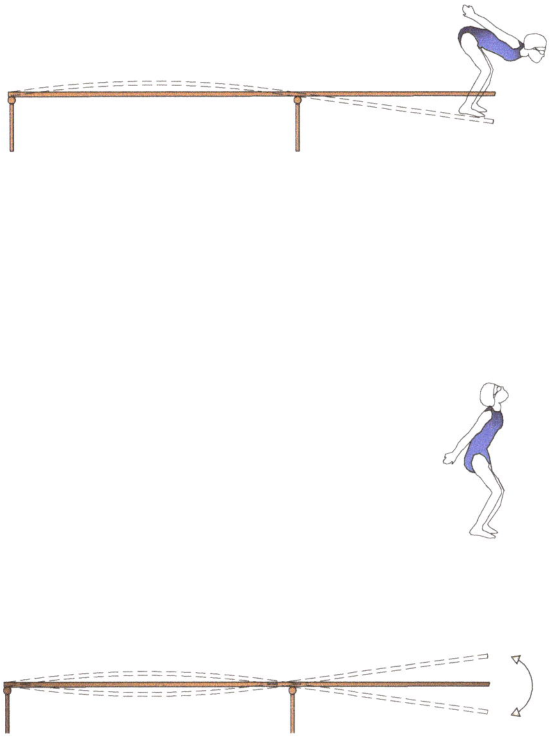

Figure 2.12.4 Dynamic Response from Dynamic Load

Resonance

The excitation from the released load is called resonance. The leaping diver in **figure 2.12.4** induces cyclical movements in the board. This sets up a resonance in the structure, which will continue until the swimmer dives into the pool. This is an unusual case where maximum excitation is desirable; a diver will tend to modify the jumping motion so that the frequency is near or matches the natural frequency of the board.

Resonance will also occur at the higher modes of natural frequency, usually with diminishing levels of response.

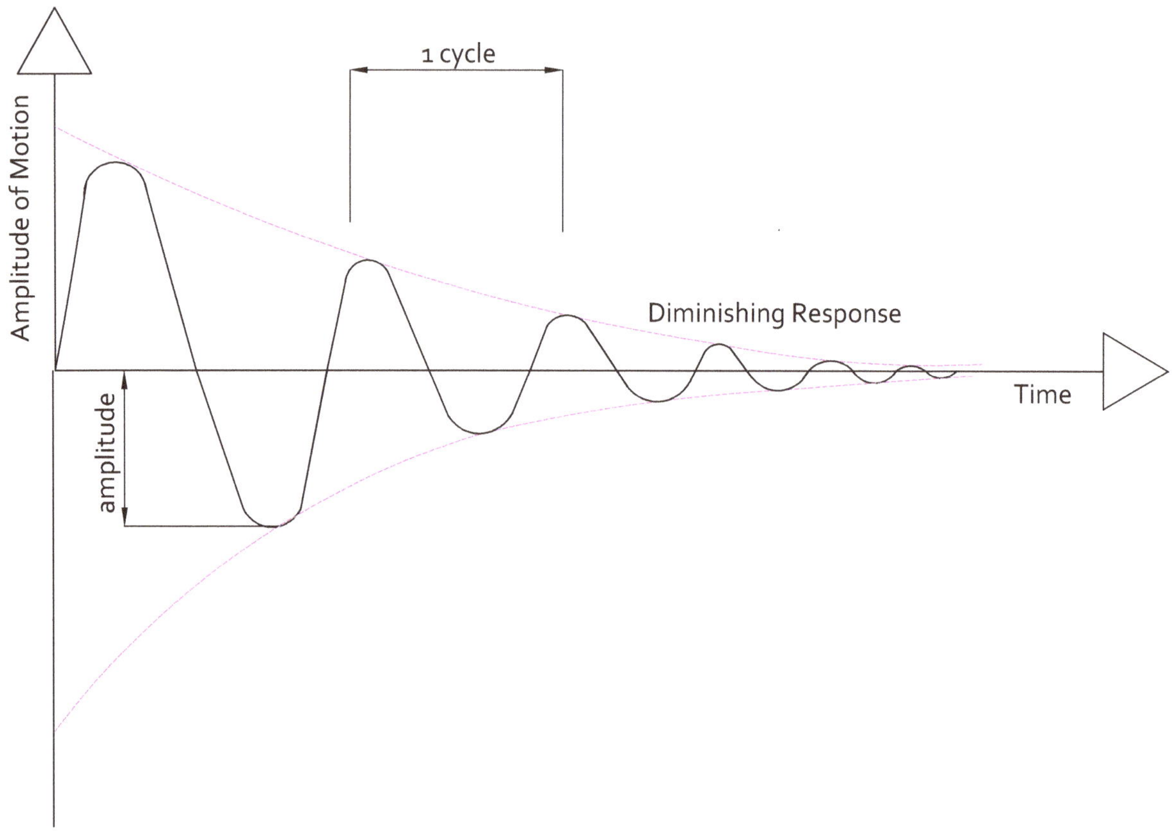

Figure 2.12.5 Damped Dynamic Motion

Acceleration of Motion

A measure of resonance is the acceleration of movement of the bridge in **figure 2.12.1** when excited by pedestrian use. If this acceleration, at resonance, increases to a level which exceeds 9.81m/ s^2 (acceleration due to gravity), the forces within the structure under its self-weight will exceed the forces which that structure will experience under normal static loading. The degree of acceleration is conveniently expressed as a *multiple of gravity*.

A 10-tonne bridge under an acceleration of 2g will experience a force equivalent to 20 tonnes. This force is a dynamic force, but to simplify the mathematics, it is considered as a *quasi-static* force response.

Damping

Damping is the phenomenon of absorption of dynamic energy, which continues until the structure is at rest: the greater the damping ratio, the faster the structure will come to rest. See **figure 2.12.5**.

All natural motion is damped. Inherent damping within a structure will derive from internal friction between structural components, such as bolted splices or internal microcracking. It is also provided by external attachments. Typical ranges of damping are shown below:

- Partitions sitting on a floor (2-6%)
- Mechanical ductwork (1-10%)
- Ceiling – suspended from floor (1-3%)
- Bare Floor (1-3%)

There is published data available for more accurate assessments from sources such as the Concrete Society.

Calculation of 1st mode Natural Frequency

For any simply supported structure, the first mode natural frequency may be calculated by working out the deflection of the structure plus a percentage of the applied dynamic or static load.

This example is based on a real structure: a 10m span ramp in a swimming pool, which was reported to be lively prior to fixing the glass balustrade. It similar to the bridge in **figure 2.12.1**. A section through the bridge is shown in **figure 2.12.6**.

For simply supported and cantilevered structures, verification of the fundamental natural frequency may be simplified using empirical formulae based on deflections under dead loading and a proportion of the design live loading.

Bridge: 1st mode of Natural Frequency

Basic Loading

Bridge structure - 2 channels along each edge, spanning 10m. Simply supported.
Bridge span = 10m
Dead Load per channel = $3.7kN/m^2$
Line Load on channel = 3.7x1.5/2
= **1.85kN/m**

Dead Load on the outer channel (half the weight of the bridge over its length).
10% of live load is generally added to the Dead Load

10% of Live Load ($3kN/m^2$)
= $0.3kN/m^2$ x 1.5/2 = 0.225kN/m

Total loading applied = 1.85+0.225 = **2.1kN/m**

Calculate Deflection under Dead + 10% Live Load

Channel moment of inertia , I = $2750cm^4$
Modulus of steel channel ,
E = 210×10^3 MPa (N/mm^2)

Deflection of simply supported channel......
$5wL^4/384EI$......

Deflection
= $5 \times 2.1 \times 10000^4/384/210 \times 10^3/27.5 \times 10^6$
= **47mm**

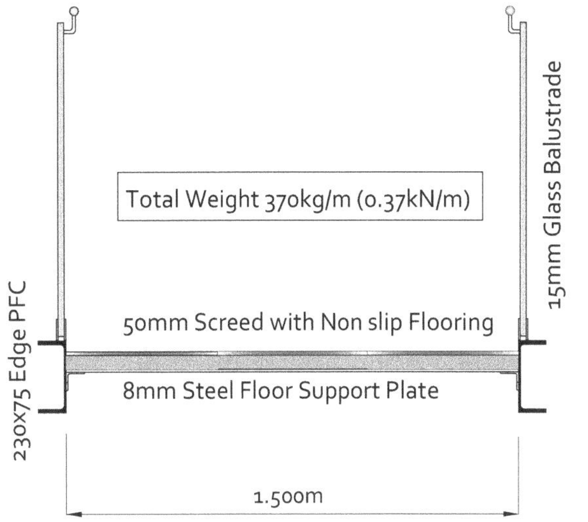

Figure 2.12.6 Section Through Footbridge

Calculate Natural Frequency of Bridge

Based on the formulae in the calculation summary which follows, from the above the natural frequency of the bridge is:

$$C = \frac{17.7}{\sqrt{\delta}} = \frac{17.7}{\sqrt{47}} = \textbf{2.6 Hz}$$

(Or 2.6 cycles per second)

2nd and 3rd modes of natural frequency may be obtained as eigenvalues from most analysis software.

A Salutary Lesson

It is instructive that this beam was designed for dead load deflection of span/250 (**50mm**). The original designer used composite steel design software. According to the software, the design was satisfactory. After the erection, the contractor reported that the bridge was very bouncy. There was a lot of concern for its use as a public swimming pool.

The glass balustrade was fixed with bolt brackets and neoprene washers. Fortunately, as soon as the balustrade was fixed, it damped the motion to a reasonable level. The bolts worked loose, but a locking compound was subsequently used to hold them in place. The bridge is in use today.

Natural Frequency Calculation

Empirical Formulae for Simple Beams

Natural Frequency = $\frac{C}{\sqrt{\delta}}$ ....units in mm

C = 15.8 cantilever point loaded

C = 19.6 cantilever with UDL (uniformly distributed load)

C = 15.8 simply supported central load

C = 17.7 simply supported UDL

Where, δ = deflection under Dead Load + 10% Live Load

Continuous Beams and Frames

For continuous beams and frames, there are no simple empirical formulae to obtain fundamental natural frequencies. 2-dimensional models are useful on computers to obtain eigenvalues (or modes of natural frequency) in this instance.

An understanding of simpler methods however is instructive when using software.

Natural Frequency of Side Sway

The simplified method described in the previous example is used, viewing the core as a vertical cantilever. This may be visualised by imagining the building tipped on its side and the weight of the building, which is the assumed modal mass, causing the static deflection, which is used to calculate the natural frequency.

Lateral 1st Natural Frequency of a Building Stabilised by a Lift Shaft

A 10 storey building which has a Dead + 10% live load per floor of **4.3kN/m²** has a **20mx20m** footprint.

See **figure 2.12.7** for an elevation of the building core and **figure 2.12.8** for a typical floor plan. The essential element to resist sideways sway is the core , assuming no stiffness contribution from the columns. This is a simplistic analysis, ignoring the fact that this configuration is not stiff torsionally.

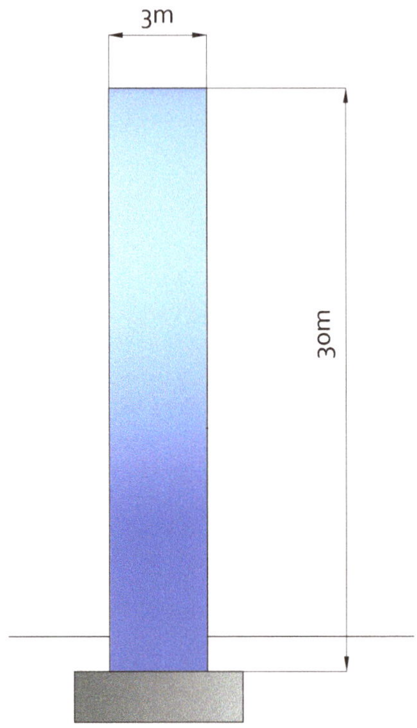

Figure 2.12.7 Section Through Core

The core comprises **200mm** thick walls and measures **3m** square on plan.

Sway Natural Frequency Calculation

In a similar manner to the previous example (**figure 2.12.6**), the deflection of the building under the calculated mass, as if turned on its side as a cantilever, is calculated.

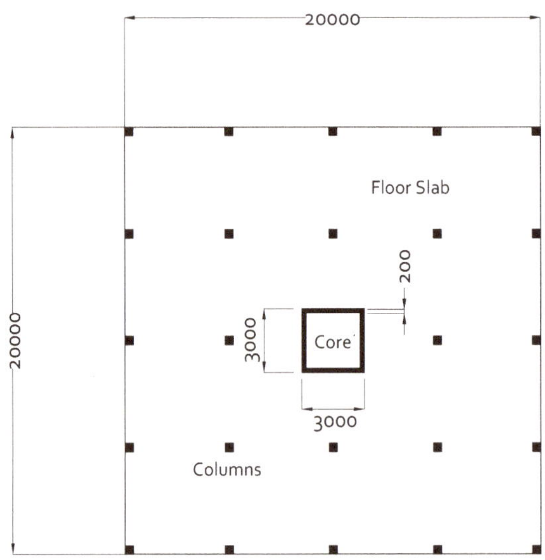

Figure 2.12.8 Building Plan

Calculate Effective Sidesway under Self-weight

Total weight of building
= 20mx20mx4.3kN/m^{2}x10 = 17200kN

Modulus of concrete (uncracked)

= 24x10^3 MPa (N/mm^2)

Moment of inertia calculated
= 3.25x10^{12} mm^4

Imagine the building as a cantilever (supported at the foundation point) and rotated perpendicular.

Deflection of building on its side
= $wL^3/8EI$ (ref. figure 2.3.1)
= 17200x10^{3}x30000^3/8/24x10^3/3.25x10^{12}
= **744mm**

From the previous example...Fn (natural frequency)
= $19.8/\sqrt{\delta}$ = $19.8/\sqrt{740}$
= **0.73 Hz**

It is to be noted that the full mass of the building was active in calculating the natural frequency. This is a gross simplification. The modal mass is the proportion of physical mass that is active in a particular mode of vibration. A detailed assessment would consider the effective modal mass in different modes of vibration, which would vary from mode to mode. A modal run of this model as a concrete core gave a modal mass of 65% of the structural mass.

Natural Period

For sway of buildings the *natural period* is often used which is the reciprocal of the natural frequency:

Natural Period of building
= 1/0.73
= **1.4 seconds**

This is a little easier to comprehend. If it resonates under wind loading it will swing to and fro in a time of 1.4 seconds.

Aside

Natural frequency is the Sum of the Parts......

A simple beam was used in the first natural frequency calculation. Where a structural system occurs, such as with a secondary girder spanning onto a primary, as in figure 2.12.9, account has to be taken of the fact that two beams (or more) are both connected.

The combined natural frequency of beam A is less than it would have been had both ends of the beam been sitting on solid supports.

The combined natural frequency of the system may be found by modelling in 3 dimensions the combination of beams and finding the deflection (due to dead + 10% live load) to obtain the natural frequency, or instructing the software to carry out a modal analysis to output eigenvalues.

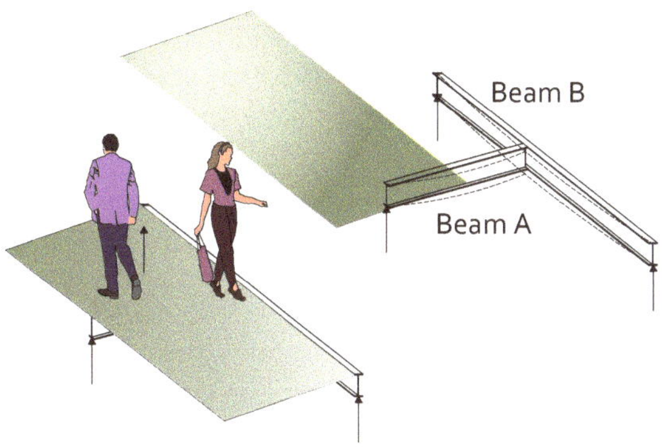

Beam A right hand support deflects (Beam B) - effective spring.

Figure 2.12.9 Secondary Beam Influence on Natural Frequency

The combined natural frequency of two beams is approximately related as follows:

$$1/fn^2 = 1/fn\ (\text{beam A})^2 + 1/fn\ (\text{beam B})^2$$

Example:

fn, both beam A and beam B is 8Hz...a desirable 'safe' frequency for each

Combined natural frequency is circa:

$$\sqrt{(1/(1/8^2 + 1/8^2)} = \mathbf{5.6\ Hz}$$

It would be necessary to increase their individual natural frequencies to 11Hz if a combined system frequency of 8 Hz is desired.

Active or Passive Mass Damping (Sway)

In Japan and New Zealand, which are earthquake zones, tall buildings may have sophisticated mass-damping systems installed at the top or bottom of the structures.

For instance, a large mass placed in a viscous fluid in a tank at the top of a building can apply very significant damping when the building is subjected to dynamic loads.

A recent example of a pendulum mass damper is contained within the Taipai 101 building in Taiwan, a 101-storey building which has a 660-tonne steel spherical pendulum suspended from the 87th to the 92nd floor. This building was designed to withstand earthquake loading and strong typhoon wind loading.

The sphere, the largest damping sphere in the world, is 5.5m in diameter and costs $ 4 million to build.

The pendulum sways when the wind blows, or earthquake tremors occur, damping the dynamic motion.

Modal Mass

Large masses on structures (for instance, a static truck on a bridge) may significantly affect the natural frequency and response.

When calculating natural frequency, any large masses should be included in the calculation in their correct position on the structure.

Structural Response of Floors

This is more difficult and requires a little more understanding of the mechanics. However, for simple floors and stadia terraces in normal use, this section is a guide to an initial assessment.

Figure 2.12.10 shows a response chart plotting dynamic magnification against the ratio of forcing frequency to natural frequency. Two plots show the response for different damping ratios (1.5% and 10%). To explain the implications of this, a number of typical responses are reviewed in the following section:

1. Forcing frequency = natural frequency

   This is the peak resonance condition.

2. Forcing frequency = 3 times natural frequency.

   Response flattens out to a point where the magnification factor of the response approaches a steady state at parity with the equivalent static response due to gravity.

3. Forcing frequency = 0.25 (25%) natural frequency

   Response becomes very low (less than the static response due to gravity).

Mitigating Factors in Real Structures

- The forcing motion (walking or similar) generally imparts a small percentage of the floor modal mass into the structure.

- Damping reduces the response at resonance.

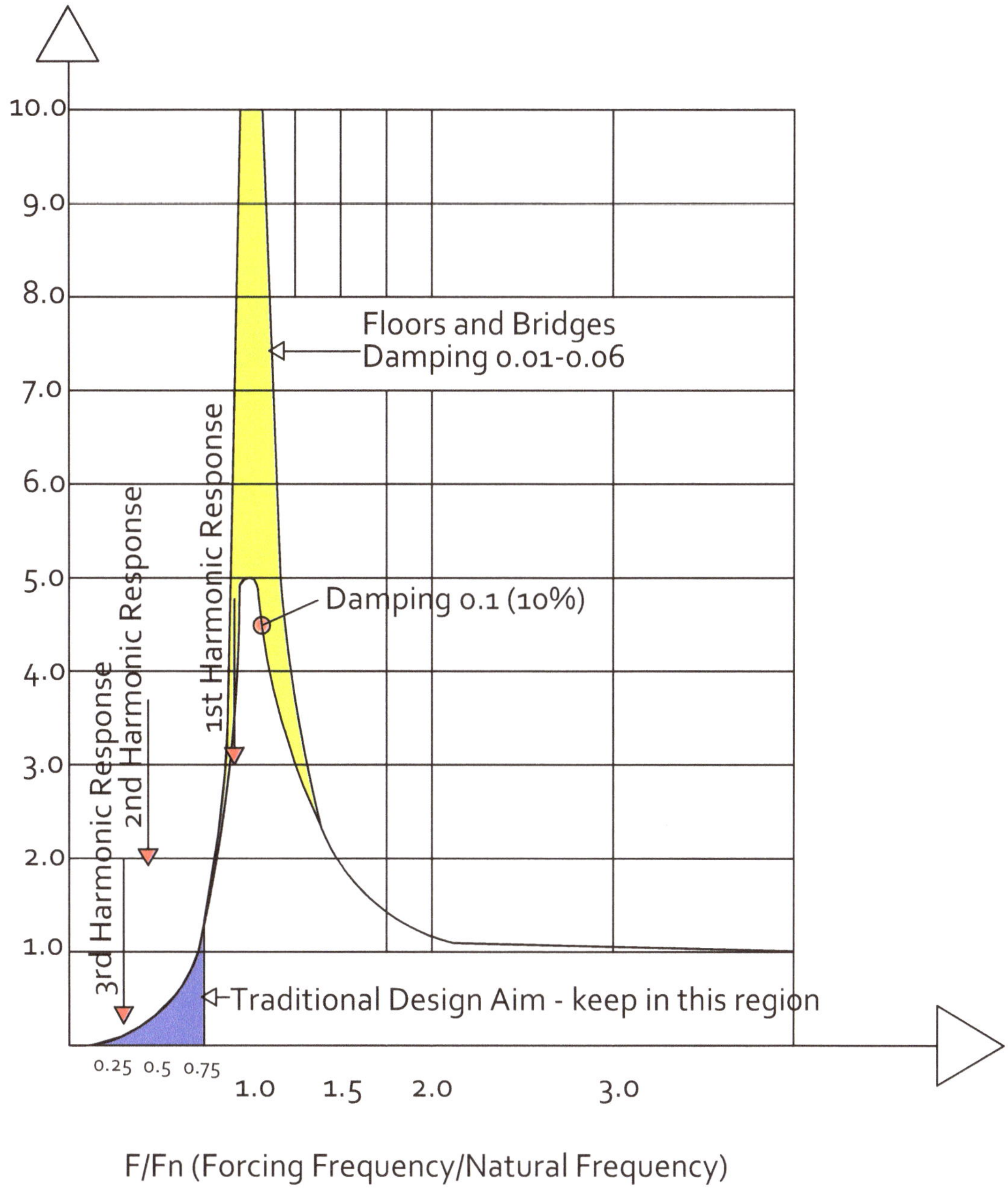

Figure 2.12.10 Response Chart

Dynamic Magnification Factor

This is the ratio of the dynamic acceleration that occurs to the static effect (i.e. under gravity). In other words, it is a measure of the dynamic force experienced by a resonating structure.

1. Forcing Frequency equals Natural Frequency (Ratio 1.0). Damping 1.5%.

Damping of 1.5% corresponds to a bare floor. If, for instance, a person walks with a forcing frequency of 2Hz and the natural frequency of the structure is also 2Hz, then resonance occurs (the yellow area of the chart). At 1.5% damping, the dynamic response is such that the magnification can become 33x the static response.

Note that increasing the damping to 3-6% will significantly reduce the response.

2. Forcing Frequency Greater than 3x Natural Frequency

The dynamic response flattens out. The mass of the system governs the response; the external force is resisted by the inertia of the system. This circumstance applies to the method of isolating reciprocating machinery. If a machine operating at 2500rpm (41 revs per second or 41Hz) sits on a heavy concrete block on isolating rubber pads with a natural frequency of 1 Hz, and this reduces the response to a minimum in the concrete block. Hence, the machinery is isolated.

Note that damping becomes irrelevant.

3. Forcing Frequency is 25% of Natural Frequency.

A structure with a natural frequency of 10Hz is walked on by a person at 2Hz. This corresponds to the blue area of the chart. The dynamic load is less than that of a static mass acting under gravity (dynamic response factor is circa 0.20). Note that damping becomes irrelevant.

Simple Formula For Assessing Structural Response

Simply ensuring that the natural frequency is in this area does not resolve the potential issues. Even if a structure is designed to have a high natural frequency, modern assessments take into account contributing factors. Based on the chart in **figure 2.12.10** the following examples are used to illustrate how response is sensitive to the interactions of *natural frequency, mass and damping*.

Peak Resonance

Where the Forcing Frequency is in the range 80% to 100% of the natural frequency. Peak resonance occurs:

$$\text{At resonance, } a/g = \frac{1.3}{2\rho} \times \frac{\alpha \cdot wp}{W}$$

Forcing Frequency greater than 120% Natural Frequency, fn

$$f/fn < 1.20,\ a/g = \frac{1.3}{((fn/f)^2-1)} \times \frac{\alpha \cdot wp}{W}$$

For dancing/rhythmic activities the multiplying factor 2.0, rather than 1.3 is advised since the 2nd and 3rd harmonic come into play.

Where........

α iis a dynamic response factor depending on the harmonic (i.e. 3rd harmonic response is less than 1st harmonic response)

wp is the load of the object which is causing the motion (around 0.75kN or 165lbs for one person))

ρ is the percentage damping. A safe assumption is usually 2-3%

fn is the natural frequency (Hz)

f is the forcing frequency (Hz) – *frequency of the crowd/person/machinery/traffic which is forcing a response in the structure under examination*

W is the weight or load of the structure which is resonating. On a floor with many bays 2-3 bays may be considered. This takes careful engineering judgment and some trial and error.

Common Forcing Frequencies from Human Activities

Indicative forcing frequencies and response factors for common activities which may be used are:

1st Harmonic

Person Walking: 1.6-2.2 Hz (α = 0.5)
Aerobics Class: 2-2.75 Hz (α = 1.5)
Dancing: 1.5-3Hz (α = 0.5)

2nd Harmonic

Person Walking: 3.2-4.4 Hz (α = 0.2)
Aerobics Class: 4-5.5 Hz (α = 0.6)

3rd Harmonic

Person walking: 4.8-6.6 (α = 0.1)
Aerobics Class: 6-8.25 (α = 0.1)

Stadium Design Loading Guide

For Stadia, spectators can act in a coordinated manner with responses (typical below):

1.5 - 3.0 Hz for hand clapping with body bouncing while swaying.

(α = 0.38 , 0.12 at 2nd harmonic, and 0.02 at 3rd harmonic).

1.5 - 3.0 Hz for a large group jumping at a rock concert.

(α = 1.5 , 0.60 at 2nd harmonic, and 0.10 at 3rd harmonic).

Examples of Estimated Responses

The calculations which follow are based on the previous formulae. As a cautionary note, they are approximations but are a good initial indicator of potential problems.

Response Factor (R) Method

Floor response criteria for office buildings and institutional buildings are often checked by more rigorous methods using *response factors*.

The Steel Construction Institute, for example, provides guidance on such methods.

Some Examples of Simple Response

Stadium Tier Response (A)

Forcing frequency less than 0.8 x Natural Frequency.

For instance, a stadium tier has a natural frequency of 3.5Hz, in the vertical mode (cantilevered stand).

The weight of the tier (as an area load) with 10% crowd load is = 8 kN/m^2

The crowd weight is $1kN/m^2$ (damping 3%) and is jumping at a frequency of 2 Hz. The crowd is standing, swaying and clapping (i.e. a = 0.38 - *see 'Stadium Design Guide' to left hand side*).

Natural frequency (3.5) is greater than 1.2xcrowd frequency (2x1.2)...

The 1st harmonic response (assuming resonance) is: $1.3 \times a \times wp / ((fn^2/f^2 - 1) \times W)$

$a/g = 1.3 \times 0.38 \times 1 / ((3.5^2/2^2 - 1) \times 8)$ = **0.03**

(i.e. an acceleration of 3%g)

Stadium Tier Response (B)

Forcing frequency equals natural frequency.

For instance, a stadium tier has a natural frequency of 2.5Hz, in the vertical mode (cantilevered stand).

The weight of the tier (as an area load) with 10% crowd load is = **8 kN/m^2**

....continued overleaf

Stadium Tier Response (B) (continued...)

The crowd weight is $1kN/m^2$ (damping 3%) and is jumping at a frequency of 2.5 Hz. The crowd is jumping to a rock concert.

Natural frequency (2.5) equals forcing frequency (2.5)...

The 1st harmonic response (assuming resonance) is: 1.3xαxwp/(2xρxW)

a/g = 1.3x1.5x1 /(2x0.03x8) = **4.1**

(i.e. an acceleration of 410%g)

Resonance and the extreme motion causes a significant response (4xg) which may also have strength implications).

Conclusion with Reference to Stadia

The Institution of Structural Engineers published recommendations (IStructE 2001, reference 6.), which noted that for spectator stands hosting sports events, vertical frequencies greater than 3.5Hz may generally be demonstrated to perform satisfactorily. Whereas, stadium structures hosting pop concerts will generally behave within acceptable limits with vertical natural frequencies greater than **6Hz** minimum.

Horizontal sway natural frequencies are also a concern. The designer should take care to consider the effects of sway adequately bracing structures to improve lateral stiffness.

The recently constructed 2012 London Olympic stadium was built with cantilever stands with natural frequencies less than 6Hz. It was demonstrated by dynamic analysis and careful engineering judgement. Economic design of stands cannot avoid dynamic analysis, which may require specialist advice.

Concrete Floor Response (Office)

Floor load = **$12kN/m^2$**, fn = **4 Hz**.

It is assumed a 5mx5m area of floor is affected (i.e. weight 300kN)

Damping 5%

Person (0.75kN) walking at 2 Hz, with a response factor, α of 0.5 gives a response:

a/g = $1.3x0.5x0.75/((4^2/2^2-1)x300)$
= **0.0001 (.010%g)** - *insignificant*

Conclusion: concrete floors tend to avoid high excitation owing to their high density (large mass).

Note: a resonance check at 2nd harmonic gave a higher peak acceleration of 0.4%g (α = 0.12)

Timber Floor Response (House)

Floor load (lightweight) = $0.7kN/m^2$,
fn = 2 Hz.

It is assumed a 5mx5m area of floor is affected (i.e. weight 17.5kN)

Damping 5%

Person (0.75kN) walking at 2 Hz, with a response factor, α of 0.5 gives a response:

Natural Frequency = forcing frequency

a/g = 1.3x0.5x0.75/(2x0.05x17.5)
= **0.28 (28%g) - very significant.**

(If the natural frequency is 8Hz...the response is (0.39%g)

Conclusion with Reference to Timber Floors

Timber floors are generally designed for natural frequencies greater than 6-8Hz. When timber joists span onto supporting girders, the designer should be aware that the supporting girder will tend to reduce the natural frequency of the system with the result that the timber joists may need to be designed with a natural frequency greater than **6- 8Hz**.

Human Perception and Comfort

Importantly, the assessment of vibration involves human comfort and perception in terms of occupied space rather than strength.

The perception of an office worker trying to concentrate on a computer screen is also going to differ markedly from the aerobics dancer or football spectator at a match. The designer must take this into account. A simplistic guide is:

0.4-0.7% g is the peak acceleration acceptable to an office worker

4-7% g is peak acceleration response acceptable to spectators of an event or aerobics participants.

Cross-reference to the example calculations is encouraged to highlight these criteria.

It has to be stressed that peak criteria are not a good measure of overall response. It is dependent on the actual frequencies. Root mean square accelerations are a more effective indicator of structural response. See below.

Root Mean Square Accelerations (RMS)

When measuring responses to structures in the field accelerometers are used. These measure real-time acceleration fluctuations, which are then post-processed as a time history to average out the peak figures. This gives a more realistic model of the dynamic behaviour of the structure.

Time History Analysis

Computer software is suited well to dynamic assessments by calculating *eigenvalues* (natural frequencies) in many modes. A more rigorous analysis may be undertaken modelling random or coordinated forcing loads in any pattern desired. This is generally termed *time history* analysis. RMS output may be obtained from this type of modelling.

Earthquake Response

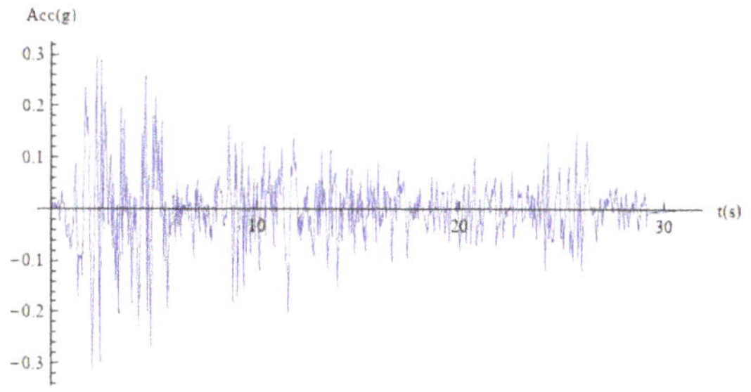

Figure 2.12.11 Ground Accelerations Plotted from an Accelerometer Plot

The response of a building to an earthquake is the imparting of energy from a laterally forced motion of the ground, which itself has a dynamic motion.

The design of a new project will involve the assessment of locally recorded data and historical evidence.

For instance, recorded peak acceleration data may be available for previous earth tremors or earthquakes. A typical response plot is shown in **figure 2.12.11**.

Analysis of a building will involve an assessment of the response under the forced vibration (in a similar way to the response of a beam under forced vibration from occupants).

Essentially (referring to **figure 2.12.7** and **2.12.8**), a building's stability-resisting core, bracing or frame may be analysed for the worst-case (peak) response using a *quasi-static* analysis: i.e. the force acting on the stabilising system is the product of the mass of the building and the peak acceleration (Force = mass x acceleration).

The worst-case structural loading in the building will generally occur on the first floor, where the building must resist the effect of the modal mass of the whole building at this level.

Horizontal Wind Sway of Tall Structures

Tall structures are susceptible to oscillations induced as a result of fluctuating wind loads, not just during storms but in the event of moderately strong winds. Wind levels are often high at upper stories of tall buildings without the *drag* effect of obstructions at ground level.

A phenomenon known as *vortex shedding* also occurs; put simply as a fluctuation of pressures which is cyclical. This is caused by the swirls of wind as the airflow is interrupted at the corners and edges of a building. The *across-wind response* is generally the most significant. Different modes of motion may occur, for instance, torsional motion or twisting.

Tall Buildings and Human Comfort

The engineering behind earthquakes and tall buildings can be complicated to understand. There are simple empirical methods which may enable the *general practitioner* to evaluate tall structures. Leonhardt, for example, suggested an empirical formula for the amplitude of motion.

Sway Motion of a Tall Building

Based on the example of **figure 2.12.7**.

Sway Motion Assessment (Leonhardt 1971, reference 7.)

Fn - natural frequency = **0.73 Hz**

Assumed amplitude of sway motion at 20m/s (45 miles per hour) wind (which is 'reasonably strong' but not a 50-100 year design gust speed).

Amplitude, estimated from above
= 0.113×20^2 = **45mm**

Check Amplitude from statics

Wind force (from standard calculation) =
$0.613\ v^2/1000$ = 0.245kPa (kN/m^2)

Total wind force on side of building
= 0.245 x 20m (width)x30m (height)
= **147kN**

Deflection under wind load as vertical cantilever, with UDL, say $WL^3/8EI$........
Deflection = $147 \times 10^3 \times (30 \times 10^3)3 / (8 \times 24 \times 10^3$ (E) $\times 3.25 \times 10^{12}$ (I)) = **6.4mm**

The peak amplitude is twice the deflection approximately (amplitude is the motion over one complete cycle with is twice the sway i.e. back and forth motion).

Amplitude = 12.8mm

Amplitude selected from the Leonhardt value of 45mm and the above of 12.8mm is the smaller value since this is more reflective of the building which, by observation, is not particularly slender.

This indicates just how careful an engineer should be in assessing sway effects. With very tall buildings greater than 15-20 stories specialist advice is strongly recommended.

Peak Sway Acceleration

A sinusoidal formula linking peak-to-peak displacement, acceleration and frequency is expressed as:

Peak Acceleration, a

$$a = 2.\pi^2.fn^2.d$$

Where.....

d = peak to peak deflection
a = acceleration in m/s^2

Figure 2.12.4......Peak Acceleration

From the chart, the amplitude is **12.8mm**, and the building sway is its natural frequency (0.73Hz).

Therefore, a $= 2 \times \pi^2 \times 0.73^2 \times 0.0128$
$= \mathbf{0.14\ m/s^2}$

This may be expressed as 1.4%g (0.014g)

Acceptance of Sway Motion

Acceptance criteria for sway motion varies according to national standards, usage of the building and the natural frequency of sway.

For example, ISO 10137 1-year return limit for residences, for the *worst 10 minutes* of sway motion, advises *peak* acceleration for a natural frequency of 0.73Hz, to be no greater than around 0.05 m/s^2. The *peak* acceleration in a **20m/s** wind gust is estimated on first assessment in the above to be **0.14m/s²** (>0.05). Further investigation is advised.

The 'Building Sickness Factor'

Awareness of lateral sway in a tall building is that peak acceleration may be sensed in the range **0.006g-0.015g** horizontally. A typical criterion is to limit peak acceleration at the top floor level to no more than 0.01g any more often than 12 times a year'.

The acceleration noted in the previous example was 0.015g, which possibly at the limit of acceptability. A higher stiffness of core might be investigated.

This design warrants further verification. A specialist may have to be consulted, but at least the simple checks have highlighted a potential problem.

Some important factors should also be noted in a more detailed assessment:

- The modal mass in the calculation was assumed to be the full weight of the floors, whereas the real modal mass is likely to be less. The natural frequency would be greater with less modal mass.
- Earthquake-prone buildings may be less susceptible to damage if the natural frequency is lower. This has been shown in buildings which have survived earthquakes in New Mexico.
- Torsional modes of vibration in a building with a single small core may be more critical.
- Stiffness of the foundations. A raft foundation on clay soil might decrease the natural frequency.

Vortex Shedding

As noted previously, vortex shedding may cause oscillations. If these oscillations are close to the building's natural frequency, then this may incur resonance.

$$fv = (Sv/b) \text{ Hz}$$

where......

fv= vortex shedding forcing frequency
v = wind speed (m/s)
b = breadth of structure across wind direction (m)

S is a constant – 0.20 circular cross section and 0.15 virtually square cross section

Note: S=0.22 (square) and 0.29 (circular) for wind speed in mph and dimensions in ft.

Taking the building from **figure 2.12.4**, which is square in plan, with a **20m/s** wind:

$$fv = 0.15 \times 20 / 20 = \mathbf{0.15 \text{ Hz}}$$

The natural frequency of the building is **0.73Hz**, which is about **4** times greater than the vortex frequency; therefore, vortex shedding may not cause a problem.

Damage to Buildings from Traffic/Pile Driving/ External Vibration

Traffic, pile driving, and blasting can be transmitted through the ground and can incur long and short-term vibration responses in a building. This is difficult to quantify. In the event of carrying out works, such as pile driving, close to structures which are anticipated to be sensitive to vibration, a good precaution is to monitor the vibration caused by the source and to stop work if a given response is exceeded.

The usual approach is to commission specialist consultants to measure vibration during sensitive operations.

Chapter 2.12

Quick Note Summary

1. Vibration in structures may be caused by occupants, machinery, traffic, plants and earthquakes.

2. The first mode of natural frequency may be readily calculated for simply supported conditions (fn – expressed in Hz or cycles per second). The 2nd and 3rd modes of vibration may also have a significant effect: they should be checked in a detailed assessment.

3. Upon release of a static load, a dynamic response occurs. Damping causes the motion to come to a standstill (like a ruler held on the edge of a table being flicked).

4. Applied dynamic forces, which are repetitive, such as walking, can set up resonance if the 1st mode of natural frequency (or upper modes) matches or is an integer multiple of the fundamental frequency of the periodical dynamic force.

5. Resonance can cause a very large energy response, of which the forces can exceed the weight of the structure under gravity. This is generally expressed as a multiple of gravity.

6. 1st mode of natural frequency may be estimated from a calculation of deflection of a structure under self weight and 10% of live loading. For continuous beams and frames, this is not as straightforward.

7. The 2nd and subsequent modes are generally calculated on software via eigenvalues.

8. Damping plays a vital role in reducing vibration response – this is the absorption of dynamic energy within a system, whether by friction or otherwise.

9. Active and passive damping methods are often employed in tall buildings subject to sway.

10. The peak response of a structure may be estimated by simple empirical equations.

11. Root mean square methods are used to form a better dynamic model by using time history peak acceleration or velocity plots. These may be obtained by computer modelling or by measuring the response of built structures with accelerometers.

12. Human perception of the response of floors is an important design criterion – in severe cases, movement may be very unpleasant or intolerable to occupants.

13. Earthquake response is determined from locally measured data – and may be expressed as a product of 'g' (gravity as acceleration) and the mass of the building elements.

14. The stabilising elements (core/bracing/walls) are assessed in terms of their resistance to the earthquake loading.

15. Tall buildings sway in the wind. The horizontal sway can be perceptible to occupants. In the worst cases,its effects are similar to sea sickness. Various methods of increasing the modal mass by linkage of the stabilising elements to outer columns have been employed, for example, floor-to-floor trusses or 'top hat' beams (Shard, London).

16. Vortex shedding of wind loads can set up oscillatory forced motion in a tall structure.

17. The effects/damage to buildings from external blast loading, traffic, construction work etc., may be monitored with sensors.

Chapter 2.13

2.13 Foundations

Any structure, from domestic to multi-storey, must be supported by the ground beneath. The interaction of geotechnics and structure cannot be avoided. Although modern construction involves geotechnical engineers almost universally, structural engineers must have a good understanding of geotechnics. This chapter reviews simple pad and strip foundations, piles, rafts and finishes with some methods of assessing retaining walls.

Foundations

Foundations are the unseen face of structural engineering, hidden from view; nevertheless an important part of structural engineering.

Foundations and substructures also provide interesting and difficult challenges, not only in selecting the 'right' solution but also in the construction sequences, which can have a major influence on project costs.

Furthermore, there are fewer 'green field' sites available in the modern world. Site investigations and assessment of buried ground obstructions are a large part of a structural consultant's workload.

Dealing with unforeseen ground problems carries a high risk for clients, engineers and contractors in terms of additional costs and construction delays.

This is a short chapter which covers some of the basic aspects of foundations and substructures.

The reader is advised to study classic texts such as 'Foundation Design and Construction' (Tomlinson 2001, reference 10.)

Simple Base Design

In its simplest form a footing or base below a wall is built as a mass concrete strip poured into a trench excavated in the ground. In domestic structures with relatively low bearing pressures, the trench is often 600mm to 900mm deep. At this level, in the UK, this level is below the depth of potentially damaging ground frosts. In colder regions, such as Siberia, with permafrost in the ground, different criteria govern.

The builder will excavate, hopefully to firm ground, whether in clay soils or sandy ground. A 2-3 storey house typical loading is customarily approaching 60kN/m .

Figure 2.13.1 shows a concrete footing. In its simplest form, the footing is made wider than the wall such that a line at 45° drawn from the outer edge of the wall intersects the outer edge of the base.

Assessing the allowable bearing pressure in the ground can necessitate site investigations and soil strength testing. A large or sloping site may have variable ground, which has to be taken into account. For the sake of this exercise, a moderately compacted formation (the bearing surface) will have an allowable bearing pressure in the region of 50-100 kPa (kN/m^2).

Theoretically, there is no bending in the base since the load is spread directly onto the ground. However, steel reinforcement is often employed, at minimum as a mesh strip, to assist in carrying and supporting the footing where bending is identified or the ground is soft.

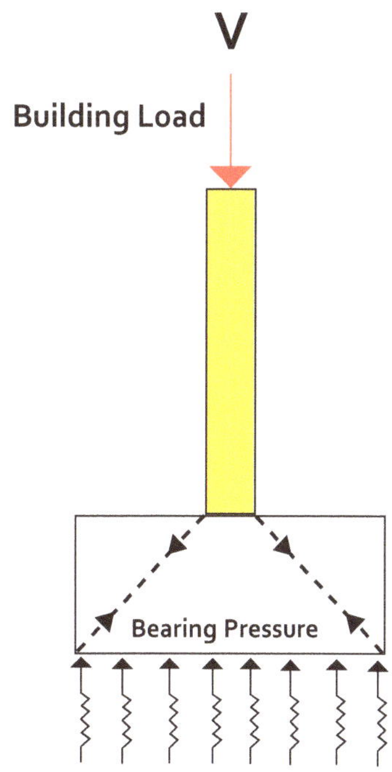

Figure 2.13.1 Pad Foundation

Ground Conditions

Ground, which is not rock, is essentially *spongy* (or based on springs), but for a reasonably stiff ground, the deformation need not be considered.

Settlement of the ground will always occur under the building loads. Where a builder has laid foundations on highly varying ground, this can lead to the problem of differential settlement.

Assuming the settlement is small, the foundation may be sized with regard to bearing pressure alone.

The example calculations below are for a wall and column base, which sit on a stiff clay with declared bearing capacities of **150** and **100kPa (150kN/m²)** respectively. See **Figure 2.13.2** and calculation below:

Wall Strip Footing

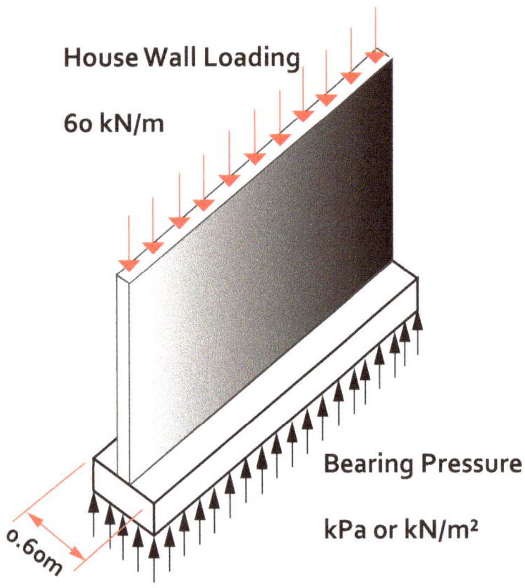

Figure 2.13.2 Wall Strip Footing

Wall footing
(600mm wide strip or linear)

Loading from house wall = 60kN/m
Bearing allowable = 150 kPa
Wall width = 300mm
Footing depth = 225mm
Foundation width = 600mm

Bearing pressure = 60/0.6 = **100kPa<150**

Note: 1kPa = 1kN/m²

The foundation is sufficient in width.

Column Footing

See **Figure 2.13.3** and calculation below:

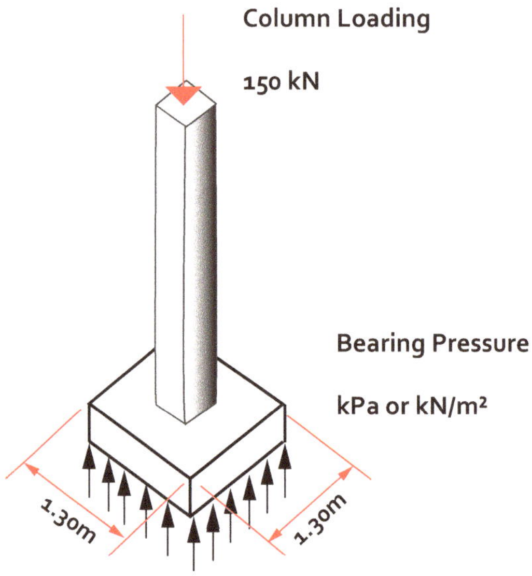

Figure 2.13.3 Pad (Column) Footing

Column Footing
(Square Base - 1.3mx1.3m on plan)

Loading from column = 150kN
Bearing allowable = 100 kPa
Column size (included baseplate) = 300x300mm
Footing depth = 500mm
Foundation dimensions = 1.3mx1.3m

Bearing pressure = 150/(1.3x1.3) = **89kPa<100**

Note: 1kPa = 1kN/m²

The column base is wide enough and deep enough at that bearing pressure.

Piled Foundations

Figure 2.13.4 overleaf shows a piled foundation. Where the ground is found to be soft or with immovable obstructions, they may be needed to transfer loads from a column or wall to a more solid substratum.

Piles may be anything from **2-3m to 50m** long, depending on the loads.

Ground Beams (or Slabs)

Piled and pad foundations are generally tied together by ground beams or a ground slab for stabilisation.

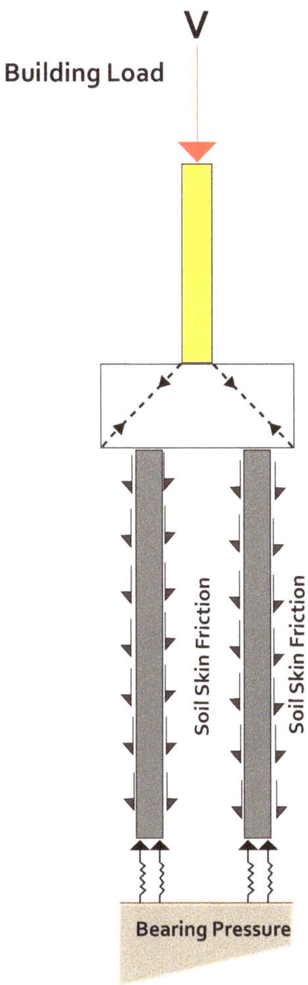

Figure 2.13.4 Piled Foundation

Pile Types

There is a large variation of pile types. Driven piles are fabricated and delivered to the site, including hollow steel tubes, H sections, interlocking sheet piles or precast concrete sections. Continuous flight auger piles (cfa) are cast in the ground. A boring rig drives forms a drill hole in the ground with a large screw under high torque; concrete is pumped into the ground as the screw withdraws. This specialist area of construction is generally undertaken by piling contractors who have the expertise to select, design and construct the piles.

Bearing Capacity of Piles

The ground bearing capacity of a pile is a combination of *friction grip* (skin friction) against the ground interface and *bearing* of the base of the pile.

A short pile of 3m may be used to support a foundation in rock, which is 2m below the surface.

In this case, the capacity is predominantly end-bearing.

A pile cap (essentially a reinforced concrete deep beam) transfers the column load or wall load and spreads it into the piles.

Piles are usually given loadings on the consultant's drawings to be used by the piling contractor to design the piles.

Pile in Clay Soil Capacity Assessment

A simple formula for approximate assessment of piles in clay soils is shown below:

Ultimate Pile Capacity

$Pu = Ab.Sub.Nc + Ap.Sua.\alpha$

where:

Ab = Area of base
Ap = Surface area of shaft

Sua = Shear strength of clay at shaft (this can vary, and can be summed in discrete lengths based on site investigation data)

Sub = undrained shear strength of clay at base of shaft

Nc = bearing coefficient (commonly 9.0)

α = adhesion factor (in the absence of data 0.45 may be used)

Safety Factor

A safety factor of 2.0 may be used to find capacity to compare with unfactored (working) loads

Typical Shear Strength Values for Soil

Hard Soil > 150 kPa (kN/mm^2)
Stiff Soil = 75-150 kPa (kN/mm^2)
Firm Soil = 40-75 kPa (kN/mm^2)

A calculation follows based on the formula presented above. This is based on simple parameters. A more comprehensive guide for piles in sandy and clay soils is found in 'Foundation Design and Construction' (Tomlinson 2001, reference 10.). The calculation overleaf gives an appreciation of a typical range of pile capacities.

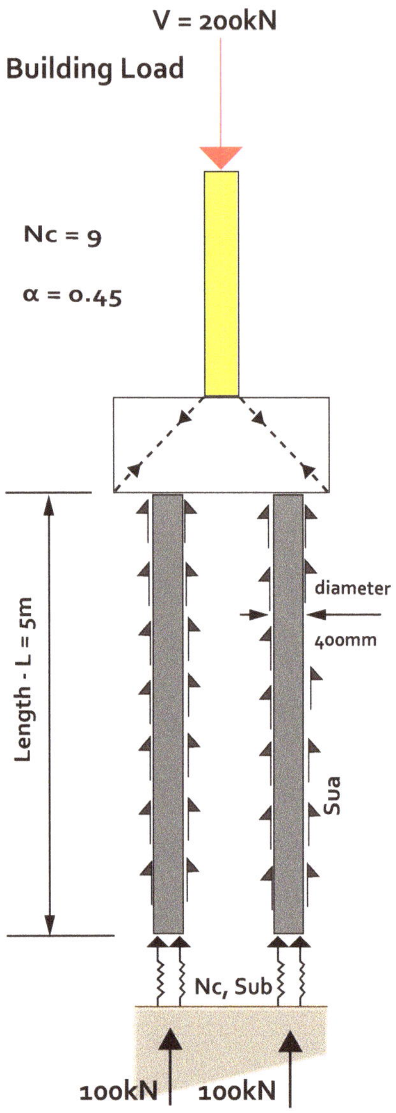

Figure 2.13.5 Pile Capacity Calculation Parameters

Pile Capacity Calculation in Clay Soil

Based on **figure 2.13.5** above, a pile capacity calculation follows. In this example, the pile cap is supported by two piles, which are assumed to be in a firm soil, with a circular precast diameter of 400mm, driven onto a hard clay, which was found in the site investigation to have a high undrained shear strength.

The parameters used in the calculation are shown in **figure 2.13.5**.

The reader is encouraged to put together a spreadsheet to test different lengths of pile, diameter and soil strengths to gain an appreciation of how soil and pile interact.

Ultimate Pile Capacity

Pu = Ab.Sub.Nc + Ap.Sua.α

where:

$Ab = \pi.\ dia.^2/4 = 3.142 \times 0.4m^2/4 = 0.126\ m^2$

$Ap = \pi.\ dia.\ L = 3.142 \times 0.4m \times 5 = 6.284m^2$

Sua = 40 kPa (kN/m^2)
Sub = 150 kPa (kN/m^2)

Nc = 9.0

α = 0.45

Calculation...

Pu = Ab.Sub.Nc + Ap.Sua.α

Pu = 0.126x150x9 + 6.284x40x0.45
= **170kN** + **113kN**
= **283kN**

Calculate working load capacity:

Safety factor of 2.0

P(w) = 283/2 = **141 kN**

Check Piles in figure 2.13.5

Applied load = **100kN per pile** < 141kN

The piles are sufficient to carry the load.

Note that the base bearing capacity (170kN) is greater than the skin friction capacity (113kN). If the pile was made longer it would penetrate the hard soil zone, with a consequent increase in capacity, and proportion of skin friction to base bearing ratio.

Raft Foundations

Sometimes, where the ground is variable in load-bearing capacity or stiffness across the footprint of a structure, the engineer might opt for a *raft*. This is essentially a stiff base which extends across the *whole footprint* and is designed to withstand the flexure or bending deformation which arises as a result of the variable ground. The reader is encouraged to read 'Foundation Design and Construction' (Tomlinson 2001, reference 10, Section 4.5) for further study.

Raft Type Examples

- Concrete rafts (useful for coping with inconsistent ground)
- Cellular rafts
- Piled rafts (a hybrid of piled and raft foundations)
- Floating basements

Retaining Walls

Retaining walls is a method of dealing with a change in level in ground. A road may be cut into a depression or on the side of a slope; the ground may be terraced on a sloping site. In both cases, a wall is built behind which the higher ground may be backfilled behind the wall.

It may be worth recapping the ring beam solution described in **chapter 2.9** at this juncture.

Repeated from the ring beam example (**figure 2.9.23**) in **chapter 2.9** retaining walls have to be designed for earth forces which push against the wall laterally.

Figure 2.9.23 is also included over the page to illustrate a retaining wall which has to retain the sides of an excavation.

Steel sheet piles or concrete piles are used to retain the ground and support the edges of the building in the long term.

Forces Acting on Retaining Walls

The wall and its footing have to resist both sliding and overturning.

Earth forces comprise (generally):

1. Lateral thrust – sideways pressure from the ground. Rather like piled grain pushing against the side of a storage bunker.

2. Surcharge – the weight of the excavating plant and traffic pushing on the ground near the excavation will exert pressure on the ground with a net effect pushing sideways on the face of the excavation.

3. Water Pressure. Sometimes, the groundwater table is at a height above the excavated level. It will exert a head of water on the side of the walls

Figure 2.13.6 illustrates a conventional brick retaining wall, which may be up to around **2m** in height. The wall may be built in a stepped arrangement and thickened at the bottom for greater height.

Figure 2.13.7 illustrates a reinforced concrete retaining wall, which is more effective for deeper retained heights or greater ground forces or surcharge. The main reinforcement is placed on the back face of the wall, which acts as a cantilever with tension forming on the back face from the leaning pressures.

Assessment of Stability Against Lateral Forces on a Retaining Wall

Figures 2.13.8, 9 and 10 on the following pages show the lateral forces that are used to check the design.

The main modes of potential instability failure which are generally considered are:

1. **Sliding Failure** – The active pressures exceeding the resistance of the wall to movement, namely the passive soil resistance at the front of the wall and frictional resistance under the wall.

2. **Overturning failure** – the overturning moments of the active pressures exceeding the restoring moment based on its self-weight and the weight of soil sitting on the back of the wall.

3. **Slip circle failure** – the whole soil mass causing a sliding failure which inscribes a circle around the whole wall. Not considered in this text, but not usually a concern for small scale retaining walls.

Sliding Failure and Overturning Calculations

Based on the parameters (above and below), are considered on **page 149**.

Slip circle analysis may be found in 'Foundation Design and Construction' (Tomlinson 2001, reference 10.)

Density of soil

A reasonable approximation is 2000kg/m^3 (20kN/m^3)

Soil Friction Coefficient

The frictional resistance of a soil mass = F.tan Ø. where F is the total vertical load on the base, Ø is the shearing resistance angle. If the wall base is smooth and precast, the resistance might reduce to F.tan (2 Ø/3).

Surcharge

The pressure of site traffic, plant etc. at the top of the wall. Typical surcharge values:

5kPa (5kN/m^2) for light traffic

10kPa (10kN/m^2) for heavy road traffic and plant
- *the minimum value used in some codes/standards*.

These are all the parameters needed for a first assessment.

Water Pressure

Water pressure increases with depth and is based on the height of the 'water column' or termed by engineers as 'the head'. *The lateral pressure of water is the same as the vertical pressure since it is a liquid.*

Water density = 1000kg/m^3 (10kN/m^3)

Soil Parameters

Soil properties are needed to undertake the calculations.

The active pressure coefficient (Ka) is based on the shearing resistance angle (Ø) of the soil; likewise, the passive pressure coefficient (Kp):

Ka = (1-sinØ)/(1+sinØ)
Kp= (1+sinØ)/(1-sinØ) .....note Kp = 1/Ka

In the absence of a site investigation a reasonable first approximation is to assume **Ø = 30°**
...giving **Ka = 0.33**

Note on Water Pressure

A good design procedure is to avoid water pressures that may build up behind the wall, not only causing saturation, but also adding considerably to the pressures acting on the wall. However, addition of water pressure if required, is easy enough to substitute into the equations.

The walls in **figures 2.13.6** and **2.13.7** have granular, self-draining soil (gravel or coarse sand) with an optional land drain (perforated tube) sitting at the back of the wall carrying water away. The brick wall has weep holes to allow water to leak out from the back of the wall.

Water pressure is measured from the top of the water table. Usually, the water table is below the top ground level. This needs to be reflected in the calculations.

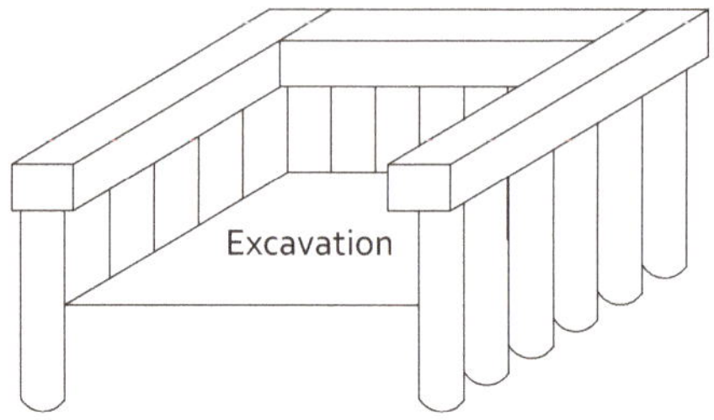

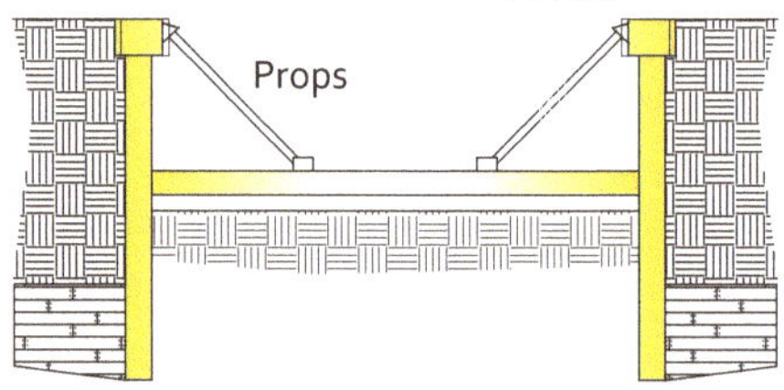

(Figure 2.9.23 Traditional Propping to a Basement) (from chapter 2.9)

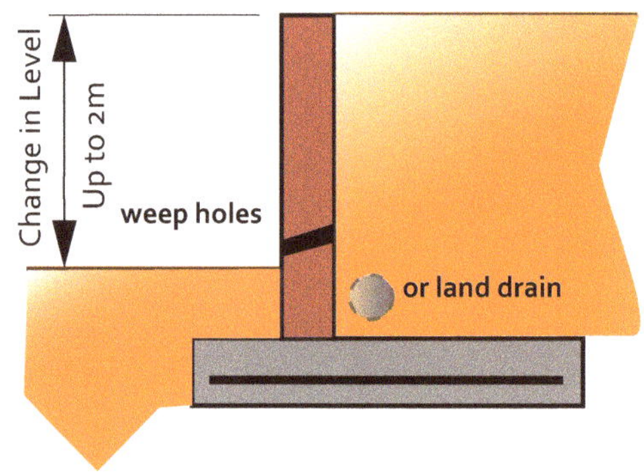

Figure 2.13.6 Solid Brick Retaining Wall

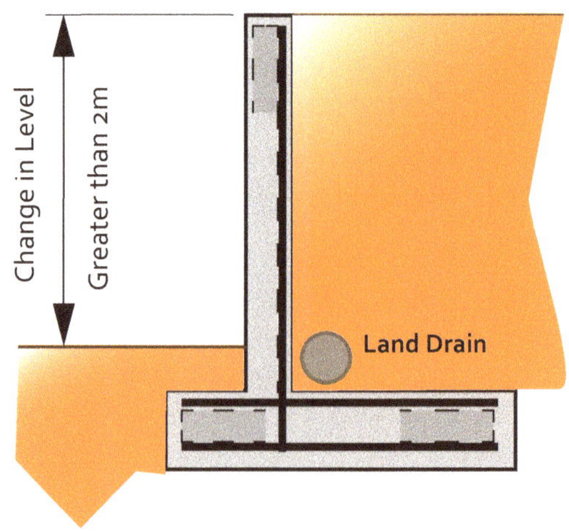

Figure 2.13.7 Reinforced Concrete Retaining Wall

Weight of Wall and Soil - a Short-cut

Most retaining wall calculations determine the weight of the wall and soil separately. However, given that the of concrete is around 20% greater than soil, an approximation will be made that the soil and concrete are one and the same.

Calculation for Sliding

See below and overleaf...

The pressure pushing behind the wall is checked against its sliding resistance.

Wall height
(soil head to underside of wall), **d** = **3m**

Depth of soil in front of wall, **h** = **0.5m**
Surcharge (surcharge pressure) = **5kPa**

Thickness of wall stem = **200mm**

Width of base behind wall = **1.5m**
Length of wall toe
(front edge to front of stem) = **0.3m**

Distance X = 0.3 + (1.5/2+0.2) = **1.25**

Angle of shearing resistance of soil, Ø = **30°**

X is essentially the 'lever arm' about which the weight of the wall acts(used in the overturning calculation).

Sliding Forces:

Density of soil = 20kN/m^3

Ka = (1-sin30°)/(1+sin30°) = 0.33

Soil pressure

= γ(soil).Ka.d^2/2
= 20 x 0.33 x 3^2/2 = **30kN/m**

Surcharge pressure
= Psurcharge.d.Ka
= 0.33x5x3 = 5kN/m

Total force acting on side of wall
= 30+5 = **35kN/m**

Sliding resistance of wall, ignoring Ppassive:

Weight of soil and wall (estimate)
= 3x(0.2+1.5)x22= **112.2kN/m**

Note: If soil density is 20kN/m^3 and concrete density is 24kN/m^3 a rough average is 22kN/m^3 as used above.

Friction resistance (***in-situ wall***):
= Ws.tan Ø
= 112.2xtan30° = **65kN/m > 35kN/m (above)**

Calculation for Overturning

This involves calculation for overturning and a resisting moment. The moments are calculated about point A in **figure 2.13.11**.

The forces calculated for sliding are used with the lever arms based on the shape of the pressure diagram; triangular is at 1/3 height and rectangular is at 1/2 height.

Overturning Moment:

Mo/t (soil) = 30kN/m x ***1/3*** x 3m =30kNm/m

Mo/t (surcharge) @ 5kPa
= 5kN/m^2 x ***1/2*** x 3m = 7.5kNm/m

Total overturning moment = **37.5kNm/m**

Restoring Moment:

Mrestoring = Weight Soil x 'X'
= 112.2kN/m x (1.7/2+.3) m
= 112.2 x 1.15 =
= **129kNm/m > 37.5kNm/m**

Aside

Note: passive resistance is ignored – this is a conservative assumption. It takes away the uncertainty of the fact that the passive resistance is reliant on the quality of compaction of the ground in front of the wall.

Summary on page 151.

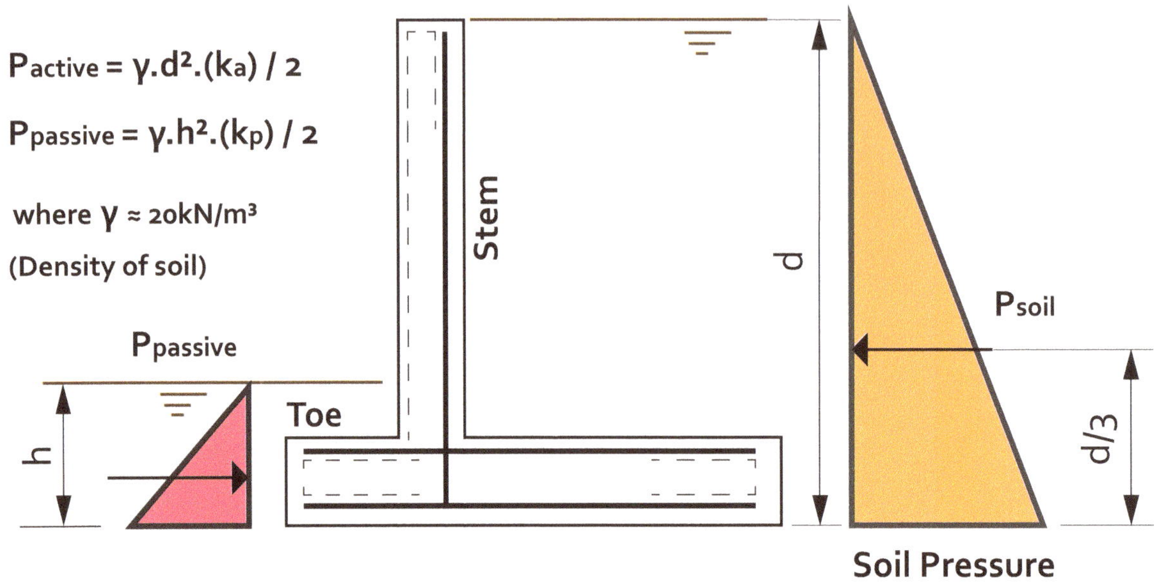

Figure 2.13.8 Retaining Wall - Soil Pressures Acting on the Wall

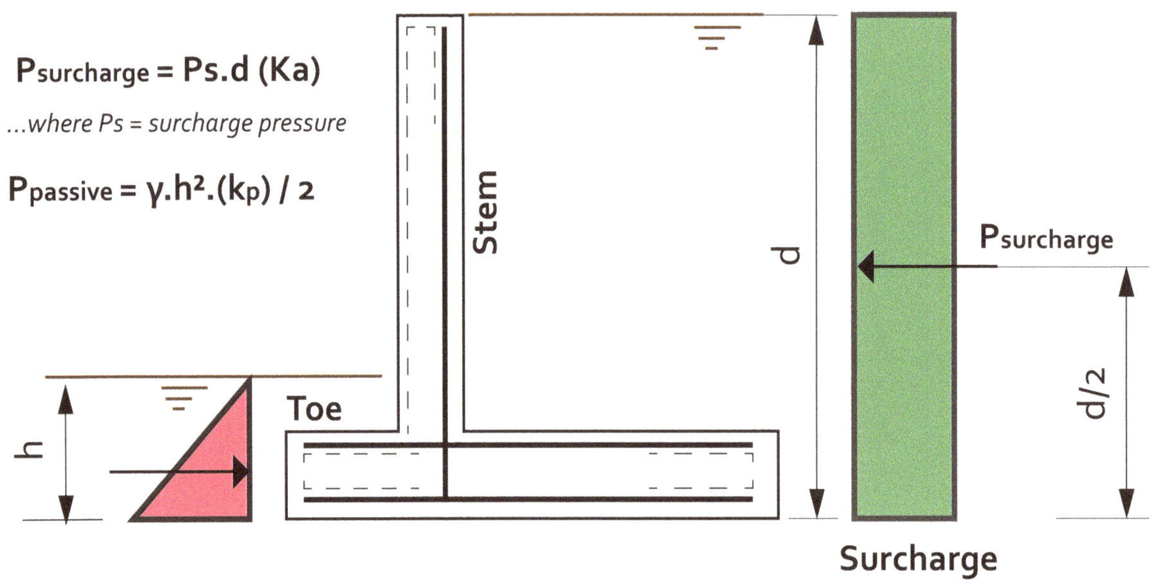

Figure 2.13.9 Retaining Wall - Surcharge Pressure Acting on the Wall

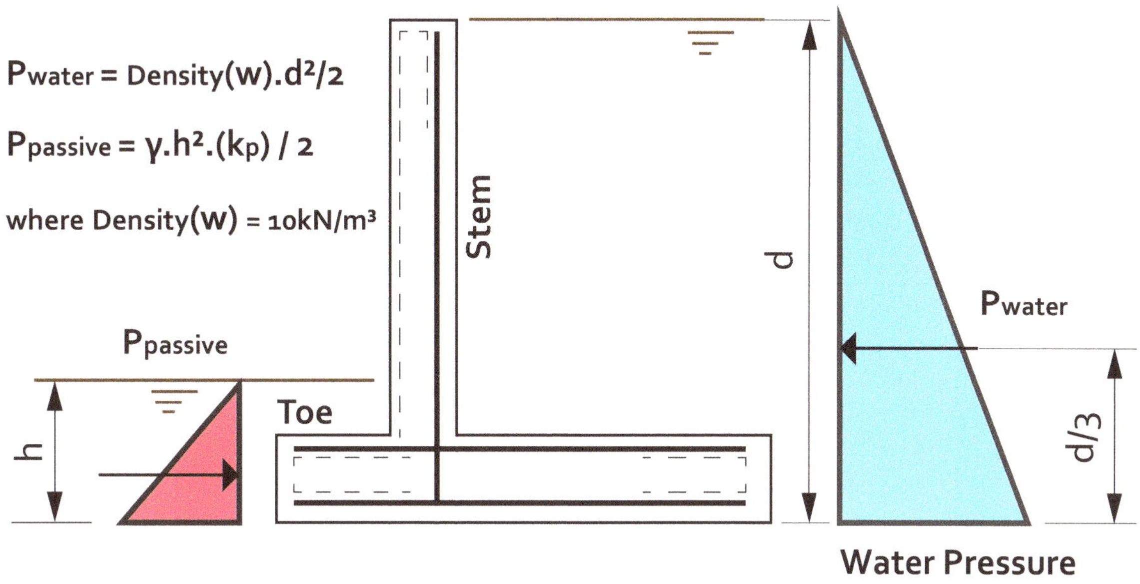

Figure 2.13.10 Retaining Wall - Water Pressure Acting on the Wall

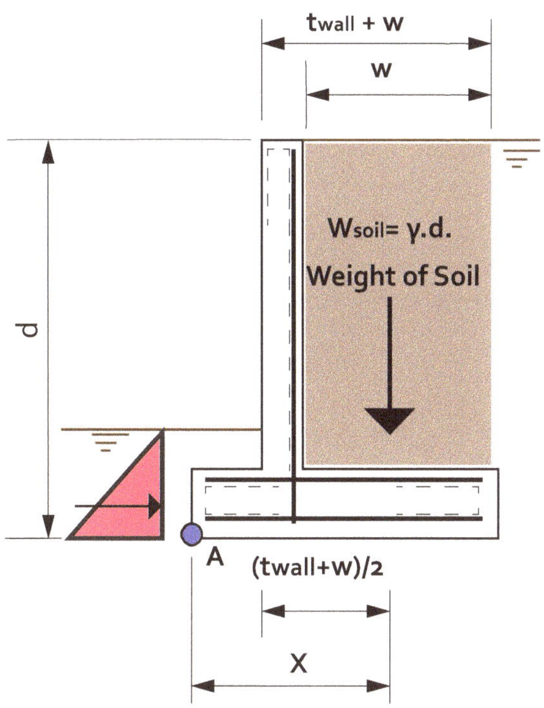

Figure 2.13.11 Soil Mass Parameters

Summary (of Page 149)

Friction resistance of wall = **65kN/m** > force acting on wall (35kN/m)

There is an effective safety factor of **1.9.** In calculating the 'stability' of equilibrium cases, it is recommended to have a safety factor. 1.50 is a common factor used.

The restoring moment, at **129kNm/m** is **3.4** times greater than the overturning moment (**37.5kNm/m**). The safety factor is well in excess of the normal safety factor circa 1.50.

The wall is stable, but any build-up of water pressure behind the wall will potentially lead to instability.

Drainage behind the wall is advisable.

Footings with Eccentric Loading

The first example of sizing a foundation with a compression loading on the base was a simple, uniform load calculation (**figures 2.13.2** & **2.13.3**).

When a bending moment is combined with an axial compression loading on a foundation, the stress distribution on the soil is no longer uniform. It is necessary to size the foundation to avoid excessive peak pressures on the base of the foundation.

Figure 2.13.11 shows a foundation with an axial load, **V**, and a bending moment, **M** acting on it.

The foundation is rectangular on plan with a breadth **B** (measured into the paper) and length **L**.

The uniform pressure of the axial load is **V/A**, where **A=BL** (**B** is the area under the base). **A** is base area.

The bending moment causes a pressure distribution which is negative (upwards) on one side and positive on the other. See **figure 2.13.13.** (**±6M/BL²**)

The pressure is found by the addition of the two resulting figures. The aggregate sum of the two is a trapezoidal or triangular shape.

In **figure 2.13.12** and **figure 2.13.13**. The eccentricity (the effective centre of action of the vertical load) is within the middle third of the foundation width. The equations and an example follow.

As long as the eccentricity is less than one-sixth of the foundation width, this equation applies. This is because the soil cannot generate tension in the base of the footing. If the eccentricity is outside the middle third, a modified pressure diagram is necessary.

Offset Bases - Reducing Eccentricity Effects

Foundations under individual columns or walls are often arranged so that the column sits 'off-centre'. For instance, in **figure 2.13.12,** if the column is moved back to the left-hand side of the base (more or less L/6 from the support), then the peak bearing pressure is reduced (the point load produces a moment opposing the applied moment).

Reinforcement Considerations

Reinforcement in pads is based on the cantilever moment and shears, lever arm from edge of column to edge of base, which may be found from the calculated bearing pressure distribution.

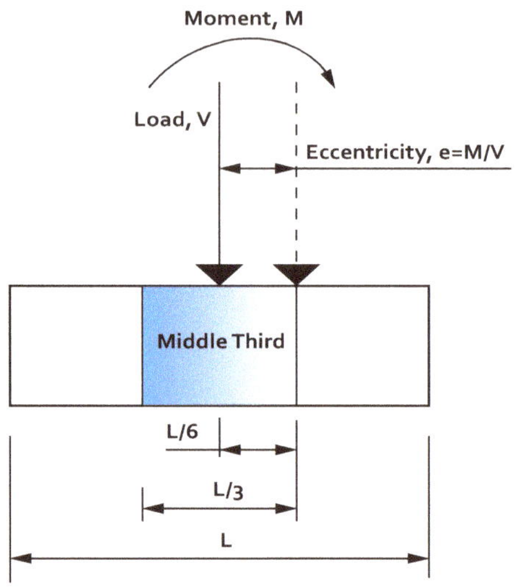

Figure 2.13.12 Eccentricity Within Middle Third

Eccentricity within Middle Third

See **figure 2.13.12.**

Where, eccentricity, $e = M/V < L/6$

When the *eccentricity* is ***inside*** the middle third, the peak pressures at the sides are found by the equation shown in **figure 2.13.13**. A calculation follows overleaf.

$$q_{max} = V/A \pm 6M/BL^2$$

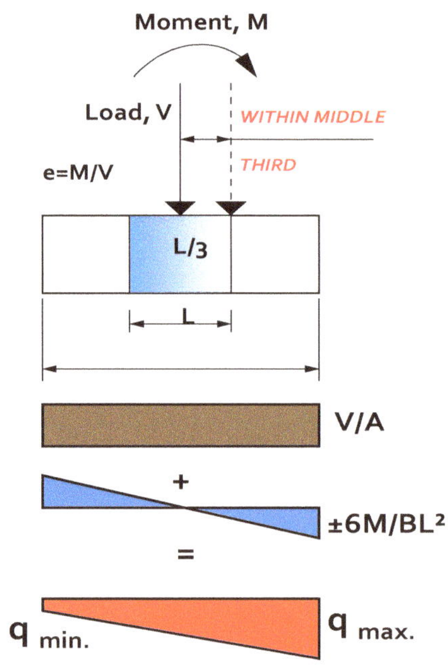

Figure 2.13.13 Eccentric Footing Pressure Distribution

Approximate Calculation of Eccentricity

Taking the retaining wall in **figure 2.13.7-10** as an example.

Bending moment
= overturning moment = **37.5kNm/m**

Weight of soil & base ~ **112.2kN/m**

Total width of base
= 0.3(toe) + 0.2(stem) +1.5m = **2.0m**

Middle Third Assessment

Now, the centre of load of the base is offset from the centre by 0.15m (1-1.7/2)...but this is ignored for approximate purposes.

Eccentricity = M/V = 37.5/112.5 = **0.33m** (left of centre)

If offset of load (0.15m) of soil and base is ignored (conservative assumption)...

Eccentricity = 0.33 **(= 0.33m - *just* within)**

The outer edge of the middle third is L/6 from centre = 0.33m.Therefore the eccentricity *just* falls within the middle third.

Pressure under base
(assuming concrete density = soil density)

Uniform pressure under load ~ 112.2/(1.0x2)
= **56kPa (kN/m²)**

Bending moment pressure = ± $M/6BD^2$
= ± $6 \times 37.5/(1.0 \times 2^2)$
= **±56kPa** (kN/m²)

Peak Bearing Pressure Check:

If the allowable bearing pressure is given as **175kPa**. Checking from **figure 2.13.12** resultant pressures are:

Maximum pressure on soil
= 56 + 56 = **112 kPa (kN/m²) < 175kPa**

Minimum pressure on soil
= 56-56 = **0 kPa (kN/m²)**....

This indicates that the base is sized within limits. A more detailed check could take into account the offset (0.15m) - see note above.

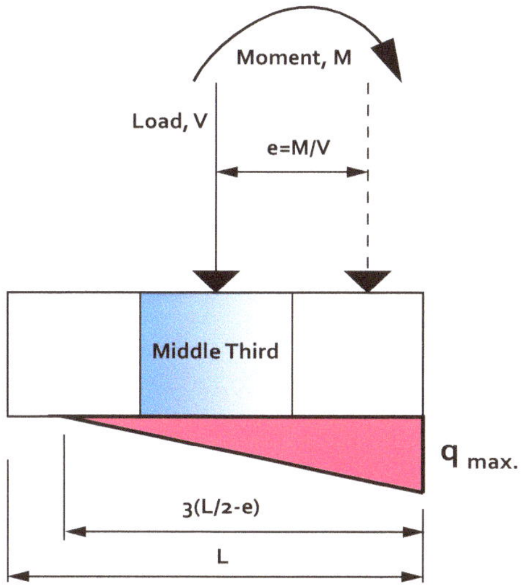

Figure 2.13.14 Eccentricity Outside Middle Third

Eccentricity Outside Middle Third

See **figure 2.13.14**.

Where, eccentricity, e = M/V > L/6.

When the *eccentricity* is ***outside*** the middle third, the peak pressure is found by the equation shown below.

qmax = 2V/3B(L/2-e)

Retaining Wall Base Example

Eccentricity Outside Middle Third

If the retaining wall in **figure 2.13.11** was involved in a flash flood for a short period, with water saturating the soil behind to a maximum level of 0.5m below ground level, there would be a change in the bearing pressure under the base - superimposed on the soil lateral pressure and surcharge - *aassessed in the calculations for sliding and overturning on the previous pages*.

Sliding resistance and overturning resistance would need to be re-evaluated, taking into account the increased lateral pressure, as illustrated in **figure 2.13.10**.

In the following pages, ignoring the sliding resistance, the increased bearing pressure is used to calculate the steel reinforcement needed in the base.

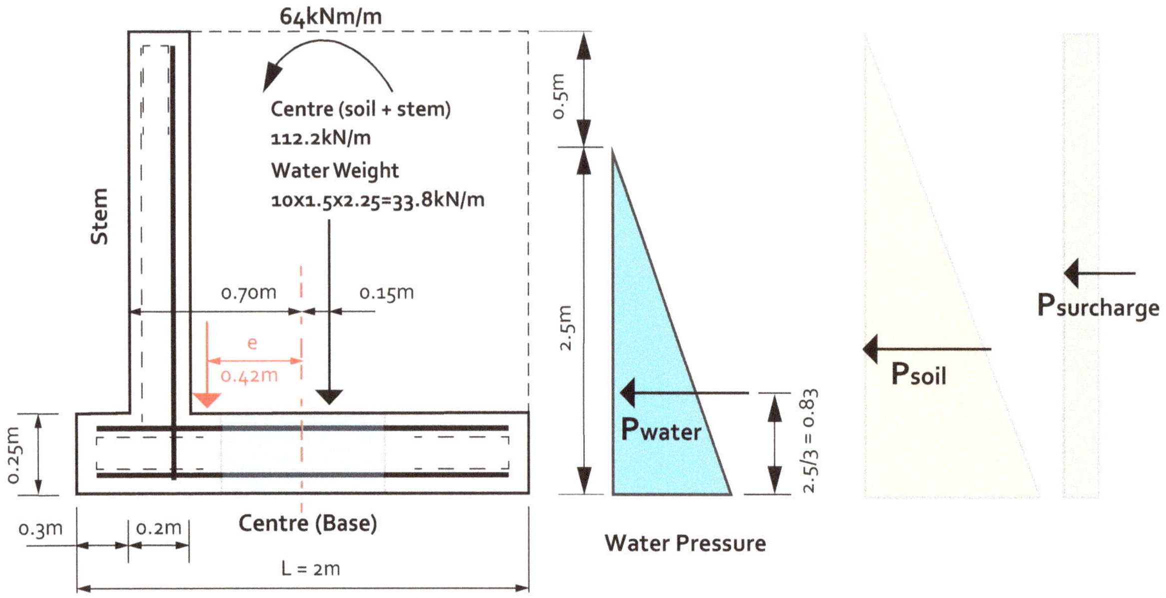

Figure 2.13.15 Flooded Retaining Wall - Eccentricity Outside Middle Third

Base Bearing Pressure Increase in Flash Flood

Figure 2.13.15 above is a summary of the lateral pressures for the retaining wall in **figure 2.13.11**. The base pressure distribution is found in the equation in **figure 2.13.14**. See **figure 2.13.15, figure 2.13.16** and below (*ignoring 1.5m centre offset*):

From figure 2.13.15....Water Pressure
=10kN/m³ x 2.5²/2 = = **31.2kN/m,**
and offset from base = 2.5m/3 = **0.83m**

Therefore 'water' overturning moment
= 31x0.83 = **26kNm/m (adding soil and surcharge moments from previous pages)**

Total overturning moment=37.5+26=**64kNm/m**

Effective Eccentricity, E = M/V
Where V = 112.2 +10x1.5x(2.5-.25) = **146kN/m**
= 64/146 = 0.43m > **0.33**
('V' includes weight of water on base)

i.e. OUTSIDE MIDDLE THIRD

Maximum pressure (from **figure 2.13.14 & 16)**

P1 (Max.) = 2V/3B (L/2-e)
= 2x146/(3x1m)/(2/2-0.43)
= **171 kPa** (N/mm²) < **175** (see previous page)

Length of pressure block = 3(L/2-e)
=3x(2/2-0.43) = 1.7m (~ **2m)**

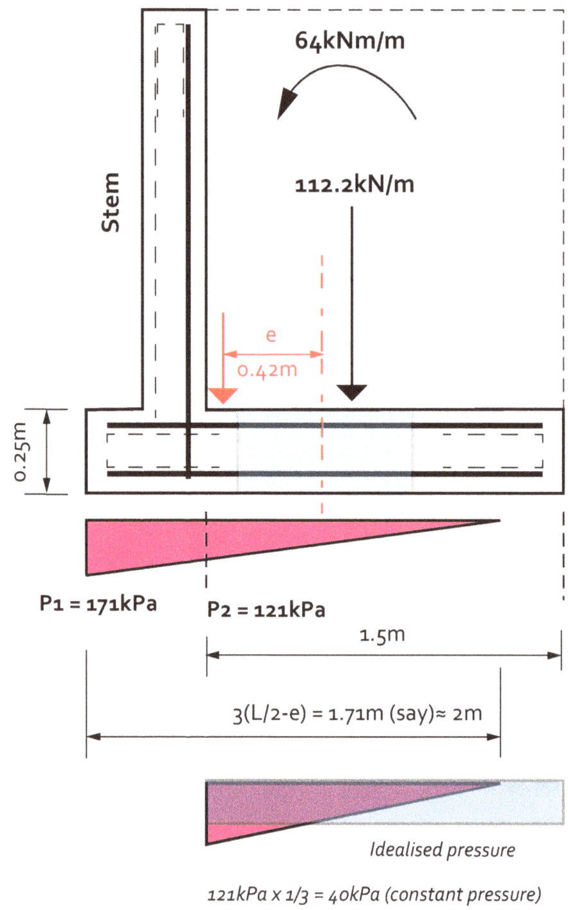

Figure 2.13.16 Idealised Base Pressure Distribution

Base Ultimate Bending Moment and Shear

An approximate calculation for reinforcement follows. This is based on an assumed plastic stress block in the base section, which is conservative.

Reinforcement area is calculated and then checked for shear based on an assumed shear strength of section for a concrete with cylinder (or cube) strength of 35-40N/mm².

See **figure 2.13.17** and calculation below:

Bending Moment and Shear in Base

Maximum bending moment occurs at inside face of stem. Cantilever bending moment (see **figure 2.3.2**). An ultimate factor of **1.5** is applied.

Idealised pressure at base from (pressure shown in **figure 2.13.16**):

P2 (stem face) = 171x(1.7-.5)/1.7 = **121kPa**

Idealised pressure (ref. **figure 2.3.9; p.26**) = 121/3 = **40kPa** (N/mm²)

Ultimate Bending Moment = 1.5xw(id.) x 1.5²/2 =1.5x40kPax1.5²/2 = **68kNm/m**

Ultimate shear force (in base critical shear is a distance of 1.5xeffective depth from the face of the stem).

Effective depth (assume 16mm diameter bars)...

$d_{eff.}$ = h(base thickness) - cover - bar dia./2 = 250mm - 50mm - 16/2 = **192mm**

1.5d = 192x1.5 = **288mm**

Ultimate shear force at 1.5d from stem face = 1.5x40kPax(1.5-0.288) = **73kN/m**

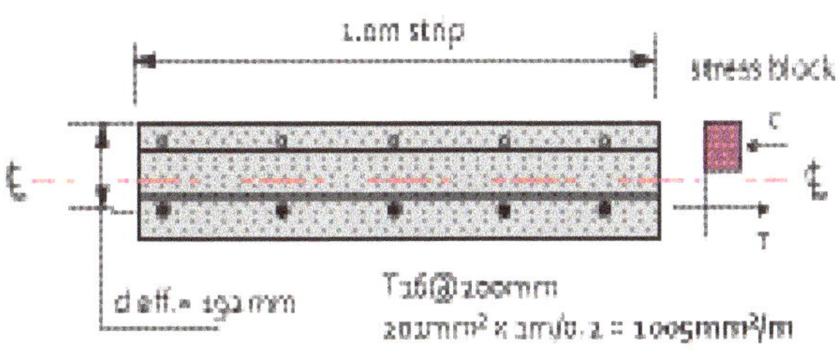

Figure 2.13.18 Base Reinforcement Section

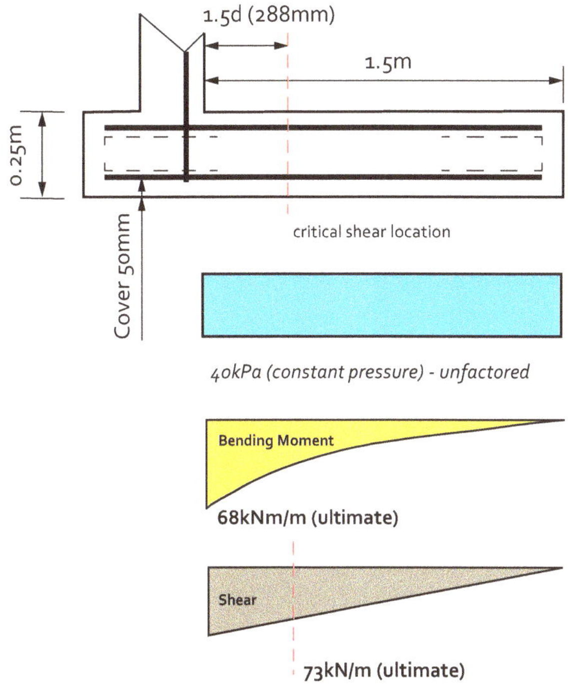

Figure 2.13.17 Base Ultimate Bending and Shear

Reinforcement Calculation

A calculation for bar area (active bars in bottom) follows. See **figure 2.13.18** and **figure 2.13.19** below:

Bar area based on figure 2.13.19:

Tension Force (T) = Mu/Lever arm = 68/0.136m = **500kN/m**

Bar tensile strength = 500 N/mm²
Area required = 500x10³/500 = 1000mm²/m

16mm bars@200mm centres - **1005mm²/m**

Shear stress = Shear/$bd_{eff.}$ = 73x10³/1000/192 = **0.38** N/mm² - Low - see **p.122** calculation

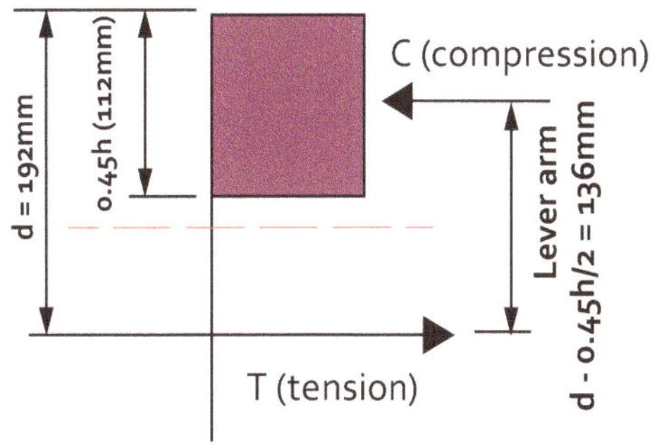

Figure 2.13.19 Base Stress Block

Chapter 2.13

Quick Note Summary

1. Foundations are effectively beams on springs. The soil stiffness may be ignored as long as a building sits on consistent, reasonably stiff ground.
2. A footing may be sized to avoid reinforcement by ensuring the load spreads directly onto the ground (450 spread).
3. Uniform pressure is generally assumed for domestic foundations.
4. Piles are used where the ground is inconsistent at a higher level. They transfer building loads into stiffer bearing strata.
5. Pile capacities are a combination of end bearing and skin friction in the shaft.
6. Retaining walls are used where a sharp change in level on a site requires holding back the ground against the higher level.
7. Retaining walls are designed to support lateral pressures: ground pressure, surcharge and water pressure.
8. Water pressure may be avoided by draining behind the wall.
9. Eccentricity is caused by a combination of overturning moment and axial loading on a foundation. A foundation is recommended to be sized so that the eccentricity falls within the middle third of the footing if possible.

Chapter 3.1

3.1 Learning from failure

"Progress, far from consisting in change, depends on retentiveness.....Those who cannot remember the past are condemned to fulfil it"

George Santayana. 19th Century - philosopher, writer.

Mistakes do Happen

Structural engineers exist to ensure that buildings stand up, that structures are fit for purpose and will perform satisfactorily over their allotted lifespan. However, humans err, and catastrophic failures have happened. There are a number of publications available on the subject of historic failures, but a highly recommended book is: 'To Engineer is Human' (Petroski 1992, reference 8.).

Many structural engineers will experience a moment of panic at the discovery of a design error, either in the calculations or the drawings.

At best, a painstaking review of calculations may reveal that assumptions made during the calculation process are over-conservative. By recalculating with the use of more accurate figures the original design is found to work. At worst, an assumed structural action may be found to be unrealistic, necessitating a design review or alteration to a detail.

Worse still, the structure may have been built. With luck, the engineer might make a small alteration in the design, return with deference to the contractor and client, admit the error and have it rectified.

This might cost the engineer or their professional indemnity insurer. Structural engineers are not given to hubris or overconfidence for good reason.

The error may have been missed entirely. The building might function with no apparent mishap until a fateful day when a terrible accident causes loss of life. A structural engineer's nightmare; it has happened before and is bound to happen again, whether through a mistake in design or for shoddy workmanship. Fortunately structural catastrophes are rare. Publicity for engineers in all disciplines comes to them in the wake of a major failure, whether it is a building collapse or a bouncy bridge.

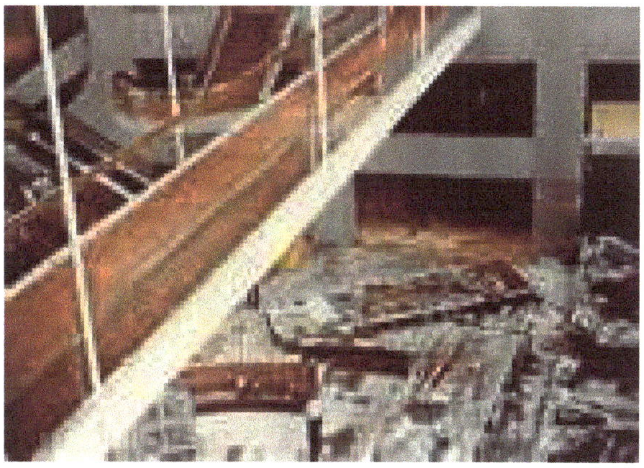

Figure 3.1.1 Hyatt Regency Walkway Collapse

Hyatt Regency Walkway Collapse

On 17 July 1981 in the Hyatt Regency hotel in Kansas City Missouri, which had been in use for two years, the internal atrium walkway collapsed during a function with sixteen hundred people present. See **figure 3.1.1**

To quote from 'Why Buildings Fall Down' (Levy and Salvadori 2002, Chapter 15, reference 9.):

'Looking up toward the source of the sound, they saw two groups of people on the second and fourth-floor walkways, observing the festivities and stomping in rhythm with the music. As the two walkways began to fall, the observers were seen holding onto the railings with terrified expressions on their faces. The fourth-floor walkway dropped from the hangers holding it to the roof structure, leaving the hangers dangling like impotent stalactites. Since the second-floor walkway hung from the fourth-floor walkway, the two began to fall together. There was a large roar as the concrete decks of the steel framed walkways cracked and crashed down, in a billowing cloud of dust, on the crowd gathered around the bar below the second-floor walkway.'

The final count of loss of life was 114, with 200 injured. The federal government commissioned an investigation by two highly respected engineers E.O.Pfrang and R.M. Marshall, who did not fully blame any party but made it clear that the structural engineers were the most culpable.

A further quotation from Salvadori and Levy concludes: 'In this design, the load of both walkways was supported every thirty feet by means of nuts screwed into a single rod on each side of the walkways at the level of the second floor and the fourth-floor box beams. Thus, the single rods hung from the steel trusses of the atrium's roof supported the weights of both walkways, but the box beams of each walkway supported only the loads of that single walkway.'

The shop drawings were produced by the steelwork contractor were based on the engineer's design, but a crucial alteration was made. The rods were offset at the fourth floor rather than running straight through as a single rod with nuts. See **figure 3.1.2**. This amendment was made on the part of the contractor so as to make the erection simpler: the uppermost fourth-floor walkway could be suspended without having to slot it through a long rod or to form a link in the rod. Once hanging, the rod for the walkway beneath could be slotted into a separate hole and the next walkway suspended. The engineers and architects approved the shop drawings.

In the interests of *buildability* (jargon for ease of construction) fatalities resulted. The rod and nut suspending the walkway floors below the fourth floor were carrying the weight of both walkways below.

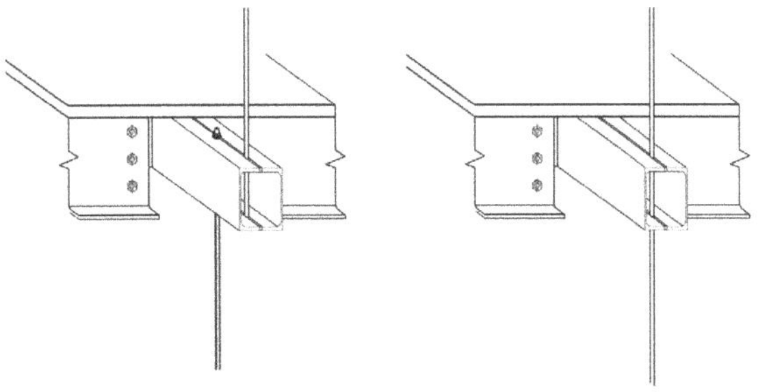

Figure 3.1.2 Hyatt Regency Walkway Detail

The rods were designed correctly for this purpose. However, the connecting beam through which the rods were threaded was transferring the full weight of the entire second floor below as well as the fourth-floor walkway.

The beam flange plates were not strong enough for this purpose. Consequently, the upper hanger located on the edge of the double channel beam pulled through the toes of the channels, the flanges having been bent and buckled out of shape.

It is of interest to pose the question: would a stricter checking regime within the engineer's design office have found the fault?

The attorney representing the state licensing board, Patrick Larney, stated: *'It wasn't a matter of doing something wrong; they just never did it at all. Nobody ever did any calculations to figure out whether or not the particular connection that held the skywalks up would work. It got built without anybody ever figuring it out if it would be strong enough. It just slipped through the cracks.'*

Many structural engineering consultancies today are, in the UK, registered as ISO 9001 quality assured (QA). Each company is audited by the regulatory body to ensure that their documents are properly filed, that the engineers are working from up-to-date design codes and that they are cross-checking calculations and drawings. In Scotland, self-certification is now a legal requirement for consultancies to check their own work. A strict auditing procedure for registered consultancies is undertaken by the 'SER' body. This takes the place of the local building control authorities, who were responsible for undertaking external design checks.

It would be an interesting exercise to circulate the Hyatt Regency walkway consultancy drawings, shop drawings and calculations to a number of quality-assured consultancies to see if they were able to spot this design error. Senior engineers are generally busy people: meeting clients, dealing with contractors, managing the design office and taking part in company management. Faced with a set of drawings and calculations, experience is necessary to read the drawings and to cross-check the calculations in an expedient manner. Some engineers prefer to pass a cursory look over the calculations during a check and to concentrate on the drawings in a speedy manner. To some extent, this is sensible. If the drawings are not clear, a red pen may be applied, and the alterations may be made before passing it on.

Concerning the Hyatt Regency detail, a sharp eyed checker would hopefully have noticed on the point at which the walkway was suspended and see that there were no stiffening plates.

Perhaps the checking engineer may then have noticed the serious implication of the contractor design change. Even at a rigorously audited QA-controlled firm, it might be easy to miss a detail like this.

If the checking engineer had designed and drawn a similar detail, perhaps they may have recommended a *more buildable* alteration based on a cross-examination of the drawings, undertaking a '*what if?*' enquiry into the details. A *visual* approach to design: conceiving details and checking details should involve the use of imagination.

A Visual Approach to Design

It is worth examining the thought process at the concept stages of a design. Moreover, it might help to explain how mistakes might be avoided.

Design is not a smooth process. For effective design, an engineer has to use an active imagination to think visually about the problems.

Imagine an engineer having to evolve a design for the node to which the fourth and second-floor walkways attach to the hangers in the Hyatt Regency Hotel discussed in the previous chapter.

The self-questioning of the designer might have followed a train of thought such as: 'What if the contractor struggles with a long rod dangling from the fourth floor walkway? What if the nut is difficult to screw all the way up 3m of rod?' See **figure 3.1.2**. It is a simple but pertinent question regarding buildability. Contractors sometimes have to alter details when presented with structures which are difficult to construct.

To re-quote Ove Arup: '*Engineering problems are under-defined; there are many solutions, good, bad and indifferent. The art is to arrive at a good solution. This is a creative activity involving imagination, intuition and deliberate choice.*' This sums up the essence of good engineering: that is, to have visualised the problems over the walkway construction to define the *under-defined* problem.

One of the solutions was the method carried out by the contractor: that is to offset the rod. Had our imaginative engineer come up with this solution, intuition might say: '*the rod dangling from the fourth floor is carrying the load of both walkways'. Would it be better to put this rod inside the second-floor rod with load-spreading washer plates to avoid flange buckling?'*

A solution to resolve the problems of erection might involve building the walkway on the ground, one storey on top of the other, with the connecting rods sized to support the temporary lightweight loading of the partially built upper platforms or with temporary cross-braced props. This whole unit might then be lifted from the permanent supporting roof beam with winches and cables; finally fixing the upper rods in place. Making reference to this on the drawings would at least have presented to the contractor a considered and safe design.

The Manchester Locks walkway detail illustrated by **figure 2.7.13** precisely solves the issues that caused the Hyatt Regency collapse. As long as the bar and threaded socket are sufficient to support the cumulated load of multiple stories of bars, and the welded connection onto the channels can transfer the load of a single level of walkway onto the tapered socket; the detail will be sufficient.

The Manchester Locks walkway caused some construction problems for the contractor with a narrow atrium and competing trades working within the building necessitating a thoroughly considered sequencing operation.

Good Engineering uses Imagination

Good engineering engages the imagination: to build up a picture, to weigh up the pros and cons of alternative solutions and to make sufficient value judgements to place them into a coherent whole.

The minutiae of design details - as illustrated by the Hyatt Regency hotel - are important. An engineering drawing is the end product of the practising engineer, whether this is expressed as a sketch on a sheet of paper or a 3-dimensional computer model. It is incumbent on each engineer to ensure that drawings directly communicate to builders, architects and clients the physical elements which are a necessary part of the assembly.

Good engineers will worry at problems until they have a clear 3 dimensional model of their structures in their heads. Might life drawing classes become a compulsory part of structural engineering courses, would this not stimulate the student to understand better *efficacy, balance and grace?* Calculations are the tools of the profession; drawings are the essential product.

Chapter 3.2

3.2 To Conclude

"It is better to be roughly right than precisely wrong."

John Maynard Keynes. 20th century economist.

One of the greatest pleasures in this profession, as it is for, say, a barrister finding the appropriate words to perform a successful cross-examination, is to be able to propose a structural solution which will resolve the problems of a perplexing project.

A structural engineer may have come up with what is thought to be the design that succeeds in solving the pertinent problems. That continuous beam solution by forming a hinge may have made the beam easier to build in shallower sections but also in a shallow section that is clear overhead and does not cause skulls to crack on impact. On first impression, an engineer may think that they have the perfect solution. All too often it fails to meet one of the criteria for the problem. It is time to rethink. Good designers always keep other options up their sleeves are always ready to seek new solutions.

It is often advisable to approach the problem quietly and pose questions to the design team at the appropriate stage: perhaps the proposal cannot be built anyway; an obstruction is discovered; the architect does not like the hinge joint, or the client wants to look at a different material.

By making suggestions and by asking searching questions before spending an imponderable amount of time drawing up the solution in detail the engineer is better informed about the physical problems involved and will have a better understanding of the architecture. Engineering is a commercial occupation: structural design can come at a cost if the initial assessment of the design does not bear any relation to the architect's proposals. Drawings are easily misread.

'It is better to be roughly right than precisely wrong' is a saying attributed to John Maynard Keynes, a 20th century economist who grappled with macroeconomic post World War II government spending and debt in the United Kingdom. This quotation is as true for engineering as it is for economics. Rough calculations and sketches at the concept stage can help enormously in understanding of the design brief. Indeed, it might be argued that they are essential for all but the most simple of problems to avoid misplaced effort, which might also lead to lost profits or even design failure.

A perennial discussion among engineers is that of their low status in society. Much of this owes to a perception in the public that engineers are pleasant boffins rather divorced from the realities of business and politics. Better communication of engineering with simplified and articulately argued designs will provide the backbone to a virtuous circle of well-managed businesses and good client relationships. It helps to improve the reputation of structural engineers.

Some engineers may take the route to management within larger organisations. Many might continue from early career to stay in touch with design and analysis. Whichever route that the prospective engineer may choose this book is intended to link the seemingly opposed worlds of mathematical analysis, practical application of design and the realm of architecture.

If the reader is a little more enthusiastic, a little more inspired and a little more enlightened as a result of reading this book, it might be judged to have been a worthwhile exercise.

Famous Engineers

John Smeaton (1724-1792) – UK
Thomas Telford (1757-1834)
George Stephenson (1780-1848)
Robert Stephenson (1803-1859)
Isambard Kingdom Brunel (1806-1859) - UK
John Roebling (1806-1869) – Austria/USA
John Fowler (1817-1898) – UK
Benjamin Baker (1840-1907) - UK
François Hennebique (1842-1921) – France
Gustave Eiffel (1832-1923)
Robert Maillart (1872-1940) – Switzerland
Owen Williams (1890-1969) – UK
Pier Luigi Nervi (1891-1979) – Italy
Fazhur Khan (1929-1982) – Bangladesh
Jean Prouvé (1901-1984) – France
Ove Arup (1995-1988) – Denmark/UK
Riccardo Morandi (1902-1989) - Italy
Peter Rice (1935-1992)
Ted Happold (1930-1996)
Felix Candela (1910-1997) – Spain
Eladio Dieste (1917-2000) – Uruguay
Frank Newby (1926-2001)
Frei Otto (1925-) – Germany
Tony Hunt (1932-) – UK
Bill Curtin (1921-1991) - UK

Some Current

Practising Engineers

Jörg Schlaich - Germany
Michel Virlogeux (1946-) -France
Chris Wise (Expedition) – UK
Stephen Morley – UK
Sam Price – UK
Paul Westbury (Buro Happold) – UK
Neil Thomas (Atelier One) - UK

References

1. Brohn D. (2005) Understanding Structural Analysis. Publisher, NewParadigm Solutions Ltd., Knightsbridge London.

2. Kennedy G.(1997) Everything is Negotiable. Publisher,The Random House Group, London. p. 76

3. Ednie J. (1925) Carpentry and Joinery. Vol. II. Publisher, London, Sir Issac Pitman and Sons Ltd.. p.102-103

4. Greenhalgh R. (1934) The Building Educator Publisher, The Gresham Publishing Company Limited. p.127

5. Heyman J. (1997) Chapter 1. The Stone Skeleton. Structural Engineering of Masonry Architecture, Publisher, Cambridge University Press, p 2.

6. The Institution of Structural Engineers, The Department for Transport, Local Government and the Regions, The Department for Culture, Media and Sport. (November 2001). Dynamic performance requirements for permanent grandstands subject to crowd action Interim guidance on assessment and design. Publisher, Institution of Structural Engineers.

7. Leonhardt F. (1971) High Rise Slender Buildings. Preliminary publication, 9th Congress LABSE, pp 213-242 (For Amsterdam Congress, May 1972)

8. Petroski H. (1992) To Engineer is Human. Publisher, Vintage Books, New York.

9. Levy M., Salvadori M. (2002). Chapter 15. Why Buildings Fall Down. Publisher, W.W.Norton and Company, New York.

10. Tomlinson M.J. (2001). Foundation Design and Construction. Publisher, Pearson Education Limited, Edinburgh Gate, Harlow, Essex, England

List of Figures By chapter

Chapter 2.1

Chapter 2.2

Chapter 2.3

Chapter 2.4

Chapter 2.5

Chapter 2.6

Chapter 2.7

Chapter 2.8

Chapter 2.9

Chapter 2.10

Chapter 2.11

Chapter 2.12

Chapter 2.13

Chapter 3.1

Full List of Photographs

(Total 32 no.) Copyright

Note – All Figures not noted below are illustrations or photographs ©Hugh Morrison or ©John Morrison.

Figure 2.1.1 Vitra Fire Station. Wikimedia Commons. GNU Free Documentation License (GNU FDL). By en: User: Sandstein

Figure 2.1.4 Peter Rice – Pavillion of the Future – Courtesy of Arup (Bruce Danzinger Los Angeles Office)

Figure 2.1.5 Sergovia Viaduct, Spain - ©John Morrison

Figure 2.2.5 Stonehenge - Wikimedia Commons. Multi-license with GFDL and Creative Commons CC-BY-SA 2.5. ByDavid Bjorgen

Figure 2.3.11 NatWest Tower (Tower 42). Flicker. Attribution Share Alike 2.0 Generic. By Salim Fadhley

Figure 2.5.7 Multistorey Frame Braced – ©Hugh Morrison

Figure 2.5.8 Aircraft Bracing –©Hugh Morrison

Figure 2.6.3 Truss Joists – ©Hugh Morrison

Figure 2.6.11 Forth of Firth – Wikimedia Commons – Estates Gazette (EG Focus) -Attribution 2.0 Generic (CC BY 2.0)

Figure 2.6.15 Infolab Bridge – ©Hugh Morrison

Figure 2.6.26 – Traditional Kingspan Roof (scan) – see reference 3

Figure 2.6.20 Vierendeel Rainhill - ©Hugh Morrison

Figure 2.6.24 Sir John Deanes College – Dramatic Overhang Entrance – ©Hugh Morrison

Figure 2.7.1 Capillano Supsension Bridge Canada – CC Creative Commons Attribution Nonderivs 2.0 Generic. By Bryan Garnett-Law

Figure 2.7.2 Severn Bridge - ©Hugh Morrison

Figure 2.7.3 Cable Stay – Hungerford Bridge – ©Hugh Morrison

Figure 2.7.8Chelsea FC South Stand Roof – ©Hugh Morrison

Figure 2.7.9 Cable Stay Entrance Infolab – ©Hugh Morrison

Figure 2.7.11 Infolab Detail – ©Hugh Morrison

Figure 2.7.12 The Locks Internal Walkway – ©Hugh Morrison

Figure 2.7.13 Tie Detail – ©Hugh Morrison

Figure 2.7.16 Goodwood Stand Roof - ©Hugh Morrison

Figure 2.8.1 Arch Bridge – Rochdale Canal Manchester – ©Hugh Morrison

Figure 2.8.3 Merchants Warehouse Manchester – ©Hugh Morrison

Figure 2.8.6 – Window arch lintels (scan) – see reference 4

Figure 2.8.12 Rome Colosseum – ©Hugh Morrison

Figure 2.8.13 Tied Brick Arch Floor (Manchester Museum of Science and Industry) - ©Hugh Morrison

Figure 2.8.15 Exeter Cathedral With Bosses. Wikimedia Commons. Public Domain. By Wanner-Laufer.

Figure 2.8.23 JFK International Terminal New York. Wikimedia Commons. Attribution 2.0 Generic (CC BY 2.0). By Pheezy

Figure 2.8.24Golpharm Stadium Huddersfield UK. Wikimedia Commons. Attribution 2.0 Generic (CC BY 2.0). By MrPenyGoff.

Figure 2.8.32 Old Trafford – ©Hugh Morrison

Figure 2.11.7 SIP Roof – ©Hugh Morrison

Figure 3.1.1 Hyatt Regency Walkway Collapse. By Dr. Lee Lowery Jr. P.E. Dr. Lee Lowery, Jr., P.E. grants anyone the right to use this work for any purpose, without any conditions, unless such conditions are required by law.

Index

Index

Index

www.ingramcontent.com/pod-product-compliance
Lightning Source LLC
LaVergne TN
LVHW071214210926
848482LV00015B/809